Steam Property Tables
Thermodynamic and Transport Properties

Based on IAPWS - IF 97
S.I. Units

by

Ashok Malhotra

Ph.D., UBC Canada, M.Tech., B.Tech., IIT Delhi

Steam Property Tables

Thermodynamic and Transport Properties

Copyright © 2012 by Ashok Malhotra.

Earlier Edition was published as:
Engineering Steam Tables
ISBN: 978 –1-4116-9935-9

ISBN-13: 978-1479230266

By CreateSpace

C ontents

---Continued on next page

INTRODUCTION

Water and steam properties are required in the power generation industry as well as numerous other engineering applications involving the flow and heat transfer of water. Mechanical engineers require these properties during the study of thermodynamics, fluid mechanics and heat transfer. Relatively small errors in property data used can result in huge losses in engineering applications such as power generation. The properties in printed form are usually referred to, as "steam tables". Nowadays, excellent computer software is available for computing steam properties. However, printed tables are nevertheless required for ready reference and in engineering examinations. The ASME has been and still is an excellent source of detailed steam tables. However, the present compact tables are more suitable for the purposes stated earlier. The present tables are a compilation of thermodynamic as well as transport properties of water. The earliest tables were due to Keenan and Keyes, 1936. An International Committee produced more accurate properties in 1967 known as IFC-67. Some thirty years later, in the nineties, the International Association for properties of Water and Steam – IAPWS - produced a new formulation, IAPWS-95, to remove known shortcomings of earlier formulations. A later modification known as the Industrial formulation 1997 (IAPWS – IF97) agrees with IAPWS-95 within acceptable engineering tolerances while giving higher computational speeds. It has now been adopted for international use. The present tables are based on these latter formulations. The primary features of the present tables are:

1. The present tables are in S.I. units. Earlier systems of units have become more or less obsolete at the present time, and were therefore not used.
2. These tables are based on IAPWS – IF97 formulations[1] that are more accurate then formulations predating the nineties for the properties of steam. Properties are listed up to a temperature of 800 ^{0}C. Higher temperature properties are available[2] but these have not been included presently as rarely required.
3. These tables are meant for engineers and students of engineering courses requiring the thermodynamic and transport properties of steam and water.
4. Properties such as h_{fg} and s_{fg} are not included since these are easily calculated as difference of other listed properties if needed. Rather the internal energy of saturated steam is included that takes more effort to calculate.
5. The tables have been arranged in easy to read clear fonts. A large size binding was used so that generous spacing and generous font sizes can be employed. The very small font sizes used in some tables cannot be read clearly by many students. These may cause errors, particularly during examinations.
6. The Mollier diagram was popular in an era predating the electronic calculator. Calculations based on it give a much lower accuracy as compared to direct use of the tables. In the author's view the Mollier diagram should be employed for conceptual understanding only, and not calculations at the present time. Therefore it has not been included along with the present tables.
7. Most of the data has been expressed using four or five significant figures. Use of larger number of decimal places is not appropriate, since it exceeds the accuracy of the formulations from which properties are computed.

---

[1] W. Wagner et al, The IAPWS Industrial Formulation 1997 for the Thermodynamic Properties of Water and Steam, Transactions of the ASME, p150, vol. 122, January 2000

[2] Ashok Malhotra and D.M.R. Panda, Thermodynamic Properties of superheated and supercritical steam, Applied Energy, P 387, Vol. 68, 2001

SYMBOLS AND UNITS USED IN TABLES

The present tables employ the following units for thermodynamic properties. Pressure has been listed in bar and temperature in degree Celsius since these are employed more frequently as compared to Mpa or Kelvin in engineering practice.

Symbol	Property	Units
P	Pressure	bar
T	Temperature	^0C
v	Specific Volume	m^3/kg
v_f	Specific Volume of Saturated Liquid	m^3/kg
v_g	Specific Volume of Saturated Vapor	m^3/kg
u	Specific Internal Energy	kJ/kg
u_f	Specific Internal Energy of Saturated Liquid	kJ/kg
u_g	Specific Internal Energy of Saturated Vapor	kJ/kg
h	Specific Enthalpy	kJ/kg
h_f	Specific Enthalpy of Saturated Liquid	kJ/kg
h_g	Specific Enthalpy of Saturated Vapor	kJ/kg
s	Specific Entropy	kJ/kg K
s_f	Specific Entropy of Saturated Liquid	kJ/kg K
s_g	Specific Entropy of Saturated Vapor	kJ/kg K

NOTE for STUDENTS: Always use lower case letters in units unless it is based on a Name. The only exception is mega where the uppercase is used for mega as in MPa .A common error is to spell quantities like kg with upper case k as in KG or KJ. Shopkeepers may do that but not Engineers.

CONVERSION FACTORS

Temperature: K = ^0C + 273.15 (*Note the Kelvin is used without a degree symbol*)
Pressure: 1 bar = 0.1 MPa = 10^5 Pa; 1 atm = 101325 Pa
To convert ft to m multiply with 0.3048
To convert ft^3 to m^3 multiply with 0.028317
1 Newton, N = 1 kg m/s^2
1 Pa = 1 N/m^2

SOME CONSTANTS

Critical Temperature of Steam = 647.096 K
Critical Pressure of Steam = 22.064 Mpa
Universal Gas Constant R = 8.31434 J/mol K (Divide this by molecular weight of a gas to get the gas constant)
For Air R = 0.287 kJ/kg K, Cpo = 1.0035kJ/kg K, Cvo = 0.7165 kJ/kg K

DRYNESS FRACTION RELATION

The properties of wet steam of a known dryness fraction, x are found with the help of the relations of the type

$$v = xv_g + (1-x) v_f$$

Replace v by the properties u, h or s in the preceding equation for other properties
Interpolation: As you will be aware, use of these tables require that you are quick and proficient at linear interpolation on your calculators. Interpolation is required when the exact pressure or temperature at which you require a property is not listed in the table but neighboring values are listed. Be sure to practice that thoroughly during your course.

Saturated Steam – Properties as function of Temperature

T ^{0}C	P, bar	v_f, m³/kg	v_g, m³/kg	u_f, kJ/kg	u_g, kJ/kg	h_f, kJ/kg	h_g, kJ/kg	s_f, kJ/kg K	s_g, kJ/kg K
0.01	0.00612	0.001000	206.0	0	2374.9	0.00	2500.9	0	9.155
1	0.00657	0.001000	192.4	4.176	2376.3	4.18	2502.7	0.0153	9.129
2	0.00706	0.001000	179.8	8.391	2377.7	8.39	2504.6	0.0306	9.103
3	0.00758	0.001000	168.0	12.60	2379.0	12.60	2506.4	0.0459	9.076
4	0.00814	0.001000	157.1	16.81	2380.4	16.81	2508.2	0.0611	9.051
5	0.00873	0.001000	147.0	21.02	2381.8	21.02	2510.1	0.0763	9.025
6	0.00935	0.001000	137.6	25.22	2383.2	25.22	2511.9	0.0913	8.999
7	0.01002	0.001000	128.9	29.42	2384.5	29.43	2513.7	0.1064	8.974
8	0.01073	0.001000	120.8	33.62	2385.9	33.63	2515.6	0.1213	8.949
9	0.01148	0.001000	113.3	37.82	2387.3	37.82	2517.4	0.1362	8.924
10	0.01228	0.001000	106.3	42.02	2388.7	42.02	2519.2	0.1511	8.900
11	0.01313	0.001000	99.79	46.21	2390.0	46.22	2521.1	0.1659	8.876
12	0.01403	0.001001	93.72	50.41	2391.4	50.41	2522.9	0.1806	8.851
13	0.01498	0.001001	88.07	54.60	2392.8	54.60	2524.7	0.1953	8.828
14	0.01599	0.001001	82.80	58.79	2394.1	58.79	2526.5	0.2099	8.804
15	0.01706	0.001001	77.88	62.98	2395.5	62.98	2528.4	0.2245	8.780
16	0.01819	0.001001	73.29	67.17	2396.9	67.17	2530.2	0.2390	8.757
17	0.01938	0.001001	69.01	71.36	2398.3	71.36	2532.0	0.2534	8.734
18	0.02065	0.001001	65.00	75.55	2399.6	75.55	2533.8	0.2678	8.711
19	0.02198	0.001002	61.26	79.73	2401.0	79.73	2535.7	0.2822	8.689
20	0.02339	0.001002	57.76	83.92	2402.4	83.92	2537.5	0.2965	8.666
21	0.02488	0.001002	54.49	88.10	2403.7	88.10	2539.3	0.3108	8.644
22	0.02645	0.001002	51.42	92.29	2405.1	92.29	2541.1	0.3250	8.622
23	0.02811	0.001003	48.55	96.47	2406.4	96.47	2542.9	0.3391	8.600
24	0.02986	0.001003	45.86	100.7	2407.8	100.66	2544.7	0.3532	8.578
25	0.03170	0.001003	43.34	104.8	2409.2	104.84	2546.5	0.3673	8.557
26	0.03364	0.001003	40.98	109.0	2410.5	109.02	2548.4	0.3813	8.535
27	0.03568	0.001004	38.76	113.2	2411.9	113.20	2550.2	0.3952	8.514
28	0.03783	0.001004	36.68	117.4	2413.2	117.38	2552.0	0.4091	8.493
29	0.04009	0.001004	34.72	121.6	2414.6	121.56	2553.8	0.4230	8.473
30	0.04247	0.001004	32.88	125.7	2415.9	125.75	2555.6	0.4368	8.452
31	0.04497	0.001005	31.15	129.9	2417.3	129.93	2557.4	0.4506	8.432
32	0.04759	0.001005	29.53	134.1	2418.7	134.11	2559.2	0.4643	8.411

T ^{0}C	P, bar	v_f, m³/kg	v_g, m³/kg	u_f, kJ/kg	u_g, kJ/kg	h_f, kJ/kg	h_g, kJ/kg	s_f, kJ/kg K	s_g, kJ/kg K
33	0.05035	0.001005	28.00	138.3	2420.0	138.29	2561.0	0.4780	8.391
34	0.05325	0.001006	26.56	142.5	2421.4	142.47	2562.8	0.4916	8.372
35	0.05629	0.001006	25.21	146.6	2422.7	146.64	2564.6	0.5052	8.352
36	0.05947	0.001006	23.93	150.8	2424.0	150.82	2566.4	0.5187	8.332
37	0.06282	0.001007	22.73	155.0	2425.4	155.00	2568.2	0.5322	8.313
38	0.06632	0.001007	21.60	159.2	2426.7	159.18	2570.0	0.5457	8.294
39	0.07000	0.001007	20.53	163.4	2428.1	163.36	2571.8	0.5591	8.275
40	0.07384	0.001008	19.52	167.5	2429.4	167.54	2573.5	0.5724	8.256
41	0.07787	0.001008	18.56	171.7	2430.8	171.72	2575.3	0.5858	8.237
42	0.08209	0.001009	17.67	175.9	2432.1	175.90	2577.1	0.5990	8.218
43	0.08650	0.001009	16.82	180.1	2433.4	180.08	2578.9	0.6123	8.200
44	0.09112	0.001009	16.01	184.2	2434.8	184.26	2580.7	0.6255	8.182
45	0.09594	0.001010	15.25	188.4	2436.1	188.44	2582.5	0.6386	8.163
46	0.1010	0.001010	14.54	192.6	2437.4	192.62	2584.2	0.6517	8.145
47	0.1063	0.001011	13.86	196.8	2438.8	196.80	2586.0	0.6648	8.128
48	0.1118	0.001011	13.21	201.0	2440.1	200.98	2587.8	0.6778	8.110
49	0.1175	0.001012	12.60	205.1	2441.4	205.16	2589.5	0.6908	8.092
50	0.1235	0.001012	12.03	209.3	2442.8	209.34	2591.3	0.7038	8.075
51	0.1298	0.001013	11.48	213.5	2444.1	213.52	2593.1	0.7167	8.058
52	0.1363	0.001013	10.96	217.7	2445.4	217.70	2594.8	0.7296	8.040
53	0.1431	0.001014	10.47	221.9	2446.7	221.88	2596.6	0.7424	8.023
54	0.1502	0.001014	10.01	226.0	2448.0	226.06	2598.4	0.7552	8.007
55	0.1576	0.001015	9.565	230.2	2449.4	230.24	2600.1	0.7680	7.990
56	0.1653	0.001015	9.145	234.4	2450.7	234.42	2601.9	0.7807	7.973
57	0.1733	0.001016	8.747	238.6	2452.0	238.61	2603.6	0.7934	7.957
58	0.1817	0.001016	8.369	242.8	2453.3	242.79	2605.4	0.8060	7.940
59	0.1904	0.001017	8.009	247.0	2454.6	246.97	2607.1	0.8186	7.924
60	0.1995	0.001017	7.668	251.1	2455.9	251.15	2608.8	0.8312	7.908
61	0.2089	0.001018	7.343	255.3	2457.2	255.34	2610.6	0.8438	7.892
62	0.2187	0.001018	7.034	259.5	2458.5	259.52	2612.3	0.8563	7.876
63	0.2288	0.001019	6.740	263.7	2459.8	263.71	2614.1	0.8687	7.861
64	0.2394	0.001019	6.460	267.9	2461.1	267.89	2615.8	0.8811	7.845
65	0.2504	0.001020	6.194	272.1	2462.4	272.08	2617.5	0.8935	7.830
66	0.2618	0.001020	5.940	276.2	2463.7	276.27	2619.2	0.9059	7.814
67	0.2737	0.001021	5.699	280.4	2465.0	280.45	2621.0	0.9182	7.799
68	0.2860	0.001022	5.468	284.6	2466.3	284.64	2622.7	0.9305	7.784
69	0.2988	0.001022	5.249	288.8	2467.6	288.83	2624.4	0.9428	7.769

T °C	P, bar	v_f, m³/kg	v_g, m³/kg	u_f, kJ/kg	u_g, kJ/kg	h_f, kJ/kg	h_g, kJ/kg	s_f, kJ/kg K	s_g, kJ/kg K
70	0.3120	0.001023	5.040	293.0	2468.9	293.02	2626.1	0.9550	7.754
71	0.3257	0.001023	4.840	297.2	2470.1	297.21	2627.8	0.9672	7.739
72	0.3400	0.001024	4.650	301.4	2471.4	301.40	2629.5	0.9793	7.725
73	0.3548	0.001025	4.468	305.6	2472.7	305.59	2631.2	0.9915	7.710
74	0.3701	0.001025	4.295	309.7	2474.0	309.78	2632.9	1.0035	7.696
75	0.3860	0.001026	4.129	313.9	2475.2	313.97	2634.6	1.0156	7.681
76	0.4024	0.001026	3.971	318.1	2476.5	318.17	2636.3	1.0276	7.667
77	0.4194	0.001027	3.820	322.3	2477.8	322.36	2638.0	1.0396	7.653
78	0.4370	0.001028	3.675	326.5	2479.0	326.56	2639.7	1.0516	7.639
79	0.4553	0.001028	3.537	330.7	2480.3	330.75	2641.3	1.0635	7.625
80	0.4741	0.001029	3.405	334.9	2481.6	334.95	2643.0	1.0754	7.611
81	0.4937	0.001030	3.279	339.1	2482.8	339.15	2644.7	1.0873	7.597
82	0.5139	0.001030	3.158	343.3	2484.1	343.34	2646.4	1.0991	7.584
83	0.5348	0.001031	3.043	347.5	2485.3	347.54	2648.0	1.1109	7.570
84	0.5564	0.001032	2.932	351.7	2486.6	351.74	2649.7	1.1227	7.557
85	0.5787	0.001032	2.826	355.9	2487.8	355.95	2651.3	1.1344	7.543
86	0.6017	0.001033	2.724	360.1	2489.0	360.15	2653.0	1.1461	7.530
87	0.6256	0.001034	2.627	364.3	2490.3	364.35	2654.6	1.1578	7.517
88	0.6502	0.001035	2.534	368.5	2491.5	368.56	2656.3	1.1694	7.504
89	0.6756	0.001035	2.445	372.7	2492.7	372.76	2657.9	1.1811	7.491
90	0.7018	0.001036	2.359	376.9	2494.0	376.97	2659.5	1.1927	7.478
91	0.7289	0.001037	2.277	381.1	2495.2	381.18	2661.2	1.2042	7.465
92	0.7568	0.001037	2.198	385.3	2496.4	385.38	2662.8	1.2158	7.453
93	0.7857	0.001038	2.123	389.5	2497.6	389.59	2664.4	1.2273	7.440
94	0.8154	0.001039	2.050	393.7	2498.8	393.81	2666.0	1.2387	7.427
95	0.8461	0.001040	1.981	397.9	2500.0	398.02	2667.6	1.2502	7.415
96	0.8777	0.001040	1.914	402.1	2501.2	402.23	2669.2	1.2616	7.403
97	0.9103	0.001041	1.850	406.4	2502.4	406.45	2670.8	1.2730	7.390
98	0.9439	0.001042	1.788	410.6	2503.6	410.66	2672.4	1.2844	7.378
99	0.9785	0.001043	1.729	414.8	2504.8	414.88	2674.0	1.2957	7.366
100	1.014	0.001043	1.672	419.0	2506.0	419.10	2675.6	1.3070	7.354
110	1.434	0.001052	1.209	461.2	2517.7	461.36	2691.1	1.4187	7.238
120	1.987	0.001060	0.8913	503.6	2528.9	503.78	2705.9	1.5278	7.129
130	2.703	0.001070	0.6681	546.1	2539.5	546.39	2720.1	1.6346	7.026
140	3.615	0.001080	0.5085	588.8	2549.6	589.20	2733.4	1.7393	6.929
150	4.761	0.001091	0.3925	631.7	2559.0	632.25	2745.9	1.8420	6.837
160	6.181	0.001102	0.3068	674.9	2567.8	675.57	2757.4	1.9428	6.749

T ^{0}C	P, bar	v_f, m^3/kg	v_g, m^3/kg	u_f, kJ/kg	u_g, kJ/kg	h_f, kJ/kg	h_g, kJ/kg	s_f, kJ/kg K	s_g, kJ/kg K
170	7.921	0.001114	0.2426	718.3	2575.7	719.21	2767.9	2.0419	6.665
180	10.026	0.001127	0.1939	762.1	2582.8	763.19	2777.2	2.1395	6.584
190	12.550	0.001141	0.1564	806.1	2589.1	807.57	2785.3	2.2358	6.506
200	15.547	0.001157	0.1272	850.6	2594.3	852.39	2792.1	2.3308	6.430
210	19.074	0.001173	0.1043	895.5	2598.4	897.73	2797.4	2.4248	6.357
220	23.193	0.001190	0.0861	940.9	2601.4	943.64	2801.1	2.5178	6.284
230	27.968	0.001209	0.0715	986.8	2603.0	990.21	2803.0	2.6102	6.213
240	33.467	0.001229	0.0597	1033.4	2603.2	1037.5	2803.1	2.7019	6.143
250	39.759	0.001252	0.0501	1080.7	2601.9	1085.7	2801.0	2.7934	6.072
260	46.921	0.001276	0.0422	1128.8	2598.8	1134.8	2796.6	2.8847	6.002
270	55.028	0.001303	0.0356	1177.9	2593.7	1185.1	2789.7	2.9762	5.930
280	64.165	0.001333	0.0302	1228.1	2586.3	1236.7	2779.8	3.0681	5.858
290	74.416	0.001366	0.0256	1279.6	2576.4	1289.8	2766.6	3.1608	5.783
300	85.877	0.001404	0.0217	1332.7	2563.5	1344.8	2749.6	3.2547	5.706
310	98.647	0.001448	0.0183	1387.7	2547.0	1402.0	2727.9	3.3506	5.624
320	112.84	0.001499	0.0155	1445.1	2526.0	1462.1	2700.7	3.4491	5.537
330	128.57	0.001561	0.0130	1505.7	2499.3	1525.7	2666.2	3.5516	5.442
340	146.00	0.001638	0.0108	1570.5	2464.6	1594.4	2622.1	3.6599	5.336
350	165.29	0.001740	0.00880	1642.1	2418.1	1670.9	2563.6	3.7783	5.211
360	186.66	0.001895	0.00694	1726.1	2351.3	1761.5	2481.0	3.9164	5.053
370	210.43	0.002222	0.00495	1845.9	2229.4	1892.6	2333.5	4.1142	4.800
373	218.13	0.002526	0.00402	1919.0	2139.8	1974.1	2227.6	4.2377	4.630

Saturated Steam - Properties as function of Pressure

P, bar	T ^{0}C	v_f, m^3/kg	v_g, m^3/kg	u_f, kJ/kg	u_g, kJ/kg	h_f, kJ/kg	h_g, kJ/kg	s_f, kJ/kg K	s_g, kJ/kg K
0.00612	0.0177	0.001000	205.9	0	2374.9	0.0332	2500.9	0	9.1553
0.007	1.88	0.001000	181.2	7.889	2377.5	7.890	2504.3	0.0288	9.1058
0.008	3.76	0.001000	159.6	15.81	2380.1	15.81	2507.8	0.0575	9.0567
0.009	5.44	0.001000	142.8	22.89	2382.4	22.89	2510.9	0.0830	9.0135
0.010	6.97	0.001000	129.2	29.30	2384.5	29.30	2513.7	0.1059	8.9749
0.011	8.37	0.001000	118.0	35.16	2386.4	35.16	2516.2	0.1268	8.9401
0.012	9.65	0.001000	108.7	40.57	2388.2	40.57	2518.6	0.1460	8.9083
0.013	10.85	0.001000	100.7	45.59	2389.8	45.59	2520.8	0.1637	8.8791
0.014	11.97	0.001001	93.9	50.28	2391.4	50.28	2522.8	0.1802	8.8521

P, bar	T ^{0}C	v_f, m^3/kg	v_g, m^3/kg	u_f, kJ/kg	u_g, kJ/kg	h_f, kJ/kg	h_g, kJ/kg	s_f, kJ/kg K	s_g, kJ/kg K
0.015	13.02	0.001001	88.0	54.68	2392.8	54.69	2524.7	0.1956	8.8270
0.016	14.01	0.001001	82.7	58.83	2394.2	58.84	2526.6	0.2101	8.8036
0.017	14.95	0.001001	78.13	62.76	2395.4	62.76	2528.3	0.2237	8.7816
0.018	15.84	0.001001	74.01	66.49	2396.7	66.49	2529.9	0.2366	8.7609
0.019	16.69	0.001001	70.32	70.04	2397.8	70.04	2531.4	0.2489	8.7413
0.02	17.50	0.001001	66.99	73.43	2398.9	73.43	2532.9	0.2606	8.7227
0.03	24.08	0.001003	45.66	100.99	2407.9	101.0	2544.9	0.3543	8.5766
0.04	28.96	0.001004	34.79	121.40	2414.5	121.4	2553.7	0.4224	8.4735
0.05	32.88	0.001005	28.19	137.76	2419.8	137.8	2560.8	0.4763	8.3939
0.06	36.16	0.001006	23.73	151.49	2424.3	151.5	2566.7	0.5209	8.3291
0.07	39.00	0.001007	20.53	163.36	2428.1	163.4	2571.8	0.5591	8.2746
0.08	41.51	0.001008	18.10	173.84	2431.4	173.9	2576.2	0.5925	8.2274
0.09	43.76	0.001009	16.20	183.25	2434.5	183.3	2580.3	0.6223	8.1859
0.10	45.81	0.001010	14.67	191.80	2437.2	191.8	2583.9	0.6492	8.1489
0.11	47.68	0.001011	13.41	199.65	2439.7	199.7	2587.2	0.6737	8.1155
0.12	49.42	0.001012	12.36	206.9	2442.0	206.9	2590.3	0.6963	8.0850
0.13	51.04	0.001013	11.46	213.7	2444.1	213.7	2593.1	0.7172	8.0570
0.14	52.55	0.001013	10.69	220.0	2446.1	220.0	2595.8	0.7366	8.0312
0.15	53.97	0.001014	10.02	225.9	2448.0	225.9	2598.3	0.7548	8.0071
0.16	55.31	0.001015	9.431	231.5	2449.8	231.6	2600.7	0.7720	7.9847
0.17	56.59	0.001015	8.909	236.9	2451.4	236.9	2602.9	0.7882	7.9636
0.18	57.80	0.001016	8.443	241.9	2453.0	241.9	2605.0	0.8035	7.9437
0.19	58.95	0.001017	8.025	246.8	2454.5	246.8	2607.0	0.8181	7.9250
0.20	60.06	0.001017	7.648	251.4	2456.0	251.4	2608.9	0.8320	7.9072
0.25	64.96	0.001020	6.203	271.9	2462.4	271.9	2617.4	0.8931	7.8302
0.30	69.10	0.001022	5.229	289.2	2467.7	289.2	2624.6	0.9439	7.7675
0.35	72.68	0.001024	4.525	304.2	2472.3	304.3	2630.7	0.9876	7.7146
0.40	75.86	0.001026	3.993	317.5	2476.3	317.6	2636.1	1.0259	7.6690
0.45	78.71	0.001028	3.576	329.5	2479.9	329.6	2640.9	1.0601	7.6288
0.50	81.32	0.001030	3.240	340.4	2483.2	340.5	2645.2	1.0910	7.5930
0.55	83.71	0.001032	2.964	350.5	2486.2	350.5	2649.2	1.1192	7.5606
0.60	85.93	0.001033	2.732	359.8	2488.9	359.8	2652.9	1.1452	7.5311
0.65	87.99	0.001035	2.535	368.5	2491.5	368.5	2656.2	1.1694	7.5040
0.70	89.93	0.001036	2.365	376.6	2493.9	376.7	2659.4	1.1919	7.4790
0.75	91.76	0.001037	2.217	384.3	2496.1	384.4	2662.4	1.2130	7.4557
0.80	93.49	0.001038	2.087	391.6	2498.2	391.6	2665.2	1.2328	7.4339
0.85	95.13	0.001040	1.972	398.5	2500.2	398.5	2667.8	1.2516	7.4135

P, bar	T °C	v_f, m³/kg	v_g, m³/kg	u_f, kJ/kg	u_g, kJ/kg	h_f, kJ/kg	h_g, kJ/kg	s_f, kJ/kg K	s_g, kJ/kg K
0.90	96.69	0.001041	1.869	405.0	2502.1	405.1	2670.3	1.2694	7.3942
0.95	98.18	0.001042	1.777	411.3	2503.8	411.4	2672.7	1.2864	7.3760
1.00	99.61	0.001043	1.694	417.3	2505.5	417.4	2674.9	1.3026	7.3588
1.05	100.98	0.001044	1.618	423.1	2507.2	423.2	2677.1	1.3180	7.3424
1.10	102.29	0.001045	1.550	428.7	2508.7	428.8	2679.2	1.3328	7.3268
1.20	104.78	0.001047	1.428	439.2	2511.6	439.3	2683.1	1.3608	7.2976
1.30	107.11	0.001049	1.325	449.0	2514.3	449.1	2686.6	1.3867	7.2708
1.40	109.29	0.001051	1.237	458.2	2516.9	458.4	2690.0	1.4109	7.2460
1.50	111.35	0.001053	1.159	466.9	2519.2	467.1	2693.1	1.4335	7.2229
1.60	113.30	0.001054	1.091	475.2	2521.4	475.3	2696.0	1.4549	7.2014
1.70	115.15	0.001056	1.031	483.0	2523.5	483.2	2698.8	1.4752	7.1811
1.80	116.91	0.001058	0.9775	490.5	2525.5	490.7	2701.4	1.4944	7.1620
1.90	118.60	0.001059	0.9293	497.6	2527.3	497.8	2703.9	1.5127	7.1440
2.00	120.21	0.001061	0.8857	504.5	2529.1	504.7	2706.2	1.5301	7.1269
2.10	121.76	0.001062	0.8462	511.1	2530.8	511.3	2708.5	1.5468	7.1106
2.20	123.25	0.001063	0.8101	517.4	2532.4	517.6	2710.6	1.5628	7.0951
2.30	124.69	0.001065	0.7771	523.5	2533.9	523.7	2712.7	1.5782	7.0802
2.40	126.07	0.001066	0.7467	529.4	2535.4	529.6	2714.6	1.5930	7.0660
2.50	127.41	0.001067	0.7187	535.1	2536.8	535.4	2716.5	1.6072	7.0524
2.60	128.71	0.001068	0.6928	540.6	2538.2	540.9	2718.3	1.6210	7.0393
2.70	129.97	0.001070	0.6687	546.0	2539.5	546.3	2720.0	1.6343	7.0267
2.80	131.19	0.001071	0.6463	551.2	2540.8	551.5	2721.7	1.6472	7.0146
2.90	132.37	0.001072	0.6254	556.2	2542.0	556.5	2723.3	1.6597	7.0029
3.00	133.53	0.001073	0.6058	561.1	2543.2	561.5	2724.9	1.6718	6.9916
3.10	134.65	0.001074	0.5874	565.9	2544.3	566.3	2726.4	1.6835	6.9806
3.20	135.74	0.001075	0.5702	570.6	2545.4	570.9	2727.9	1.6950	6.9700
3.30	136.81	0.001076	0.5540	575.1	2546.5	575.5	2729.3	1.7061	6.9597
3.40	137.85	0.001078	0.5387	579.6	2547.5	580.0	2730.6	1.7169	6.9498
3.50	138.86	0.001079	0.5242	583.9	2548.5	584.3	2732.0	1.7275	6.9401
3.60	139.85	0.001080	0.5105	588.2	2549.5	588.6	2733.3	1.7378	6.9307
3.70	140.82	0.001081	0.4975	592.3	2550.4	592.7	2734.5	1.7478	6.9215
3.80	141.77	0.001082	0.4852	596.4	2551.3	596.8	2735.7	1.7576	6.9126
3.90	142.70	0.001083	0.4735	600.4	2552.2	600.8	2736.9	1.7672	6.9039
4.00	143.61	0.001084	0.4624	604.3	2553.1	604.7	2738.1	1.7766	6.8954
4.10	144.50	0.001085	0.4518	608.1	2553.9	608.6	2739.2	1.7858	6.8872
4.20	145.38	0.001085	0.4417	611.9	2554.8	612.3	2740.3	1.7948	6.8791
4.30	146.24	0.001086	0.4320	615.6	2555.6	616.0	2741.3	1.8036	6.8712

P, bar	T ^{0}C	v_f, m³/kg	v_g, m³/kg	u_f, kJ/kg	u_g, kJ/kg	h_f, kJ/kg	h_g, kJ/kg	s_f, kJ/kg K	s_g, kJ/kg K
4.40	147.08	0.001087	0.4227	619.2	2556.4	619.7	2742.4	1.8122	6.8635
4.50	147.91	0.001088	0.4139	622.7	2557.1	623.2	2743.4	1.8206	6.8560
4.60	148.72	0.001089	0.4054	626.2	2557.9	626.7	2744.4	1.8289	6.8486
4.70	149.52	0.001090	0.3973	629.7	2558.6	630.2	2745.3	1.8371	6.8414
4.80	150.30	0.001091	0.3895	633.0	2559.3	633.6	2746.3	1.8450	6.8343
4.90	151.08	0.001092	0.3820	636.4	2560.0	636.9	2747.2	1.8529	6.8274
5.00	151.84	0.001093	0.3748	639.6	2560.7	640.2	2748.1	1.8606	6.8206
5.10	152.58	0.001093	0.3679	642.9	2561.4	643.4	2749.0	1.8682	6.8139
5.20	153.32	0.001094	0.3612	646.0	2562.0	646.6	2749.9	1.8756	6.8074
5.30	154.04	0.001095	0.3548	649.2	2562.7	649.7	2750.7	1.8829	6.8010
5.40	154.76	0.001096	0.3486	652.2	2563.3	652.8	2751.5	1.8901	6.7947
5.50	155.46	0.001097	0.3426	655.3	2563.9	655.9	2752.3	1.8972	6.7885
5.60	156.15	0.001097	0.3368	658.3	2564.5	658.9	2753.1	1.9042	6.7824
5.70	156.84	0.001098	0.3312	661.2	2565.1	661.8	2753.9	1.9111	6.7765
5.80	157.51	0.001099	0.3258	664.1	2565.7	664.8	2754.7	1.9179	6.7706
5.90	158.18	0.001100	0.3206	667.0	2566.2	667.7	2755.4	1.9245	6.7648
6.00	158.83	0.001101	0.3156	669.8	2566.8	670.5	2756.1	1.9311	6.7592
6.10	159.48	0.001101	0.3107	672.6	2567.3	673.3	2756.9	1.9376	6.7536
6.20	160.12	0.001102	0.3059	675.4	2567.9	676.1	2757.6	1.9440	6.7481
6.30	160.75	0.001103	0.3014	678.1	2568.4	678.8	2758.3	1.9503	6.7427
6.40	161.37	0.001104	0.2969	680.8	2568.9	681.5	2758.9	1.9565	6.7374
6.50	161.99	0.001104	0.2926	683.5	2569.4	684.2	2759.6	1.9626	6.7321
6.60	162.59	0.001105	0.2884	686.1	2569.9	686.9	2760.2	1.9686	6.7269
6.70	163.19	0.001106	0.2843	688.7	2570.4	689.5	2760.9	1.9746	6.7218
6.80	163.79	0.001107	0.2804	691.3	2570.9	692.1	2761.5	1.9805	6.7168
6.90	164.37	0.001107	0.2765	693.9	2571.4	694.6	2762.1	1.9863	6.7119
7.00	164.95	0.001108	0.2728	696.4	2571.8	697.1	2762.7	1.9921	6.7070
7.10	165.53	0.001109	0.2691	698.9	2572.3	699.6	2763.3	1.9978	6.7022
7.20	166.09	0.001109	0.2656	701.3	2572.7	702.1	2763.9	2.0034	6.6974
7.30	166.65	0.001110	0.2621	703.8	2573.2	704.6	2764.5	2.0089	6.6927
7.40	167.21	0.001111	0.2588	706.2	2573.6	707.0	2765.1	2.0144	6.6881
7.50	167.76	0.001111	0.2555	708.6	2574.0	709.4	2765.6	2.0198	6.6835
7.60	168.30	0.001112	0.2523	710.9	2574.4	711.8	2766.2	2.0252	6.6790
7.70	168.83	0.001113	0.2492	713.2	2574.8	714.1	2766.7	2.0305	6.6746
7.80	169.37	0.001113	0.2462	715.6	2575.2	716.4	2767.3	2.0357	6.6702
7.90	169.89	0.001114	0.2432	717.9	2575.6	718.7	2767.8	2.0409	6.6658
8.00	170.41	0.001115	0.2403	720.1	2576.0	721.0	2768.3	2.0460	6.6615

P, bar	T °C	v_f, m³/kg	v_g, m³/kg	u_f, kJ/kg	u_g, kJ/kg	h_f, kJ/kg	h_g, kJ/kg	s_f, kJ/kg K	s_g, kJ/kg K
8.10	170.93	0.001115	0.2375	722.4	2576.4	723.3	2768.8	2.0511	6.6573
8.20	171.44	0.001116	0.2348	724.6	2576.8	725.5	2769.3	2.0561	6.6531
8.30	171.95	0.001117	0.2321	726.8	2577.2	727.7	2769.8	2.0610	6.6490
8.40	172.45	0.001117	0.2294	729.0	2577.6	729.9	2770.3	2.0659	6.6449
8.50	172.94	0.001118	0.2269	731.2	2577.9	732.1	2770.8	2.0708	6.6408
8.60	173.43	0.001119	0.2244	733.3	2578.3	734.3	2771.2	2.0756	6.6368
8.70	173.92	0.001119	0.2219	735.4	2578.6	736.4	2771.7	2.0804	6.6329
8.80	174.41	0.001120	0.2195	737.5	2579.0	738.5	2772.1	2.0851	6.6289
8.90	174.88	0.001121	0.2172	739.6	2579.3	740.6	2772.6	2.0898	6.6251
9.00	175.36	0.001121	0.2149	741.7	2579.7	742.7	2773.0	2.0944	6.6212
9.10	175.83	0.001122	0.2126	743.8	2580.0	744.8	2773.5	2.0990	6.6174
9.20	176.29	0.001122	0.2104	745.8	2580.3	746.8	2773.9	2.1035	6.6137
9.30	176.76	0.001123	0.2083	747.8	2580.6	748.9	2774.3	2.1080	6.6100
9.40	177.21	0.001124	0.2062	749.8	2581.0	750.9	2774.7	2.1125	6.6063
9.50	177.67	0.001124	0.2041	751.8	2581.3	752.9	2775.2	2.1169	6.6027
9.60	178.12	0.001125	0.2021	753.8	2581.6	754.9	2775.6	2.1213	6.5991
9.70	178.57	0.001125	0.2001	755.8	2581.9	756.9	2776.0	2.1256	6.5955
9.80	179.01	0.001126	0.1981	757.7	2582.2	758.8	2776.3	2.1299	6.5919
9.90	179.45	0.001127	0.1962	759.6	2582.5	760.8	2776.7	2.1342	6.5884
10.00	179.89	0.001127	0.1943	761.6	2582.8	762.7	2777.1	2.1384	6.5850
11.00	184.07	0.001133	0.1774	780.0	2585.5	781.2	2780.7	2.1789	6.5520
12.00	187.96	0.001139	0.1632	797.1	2587.9	798.5	2783.8	2.2163	6.5217
13.00	191.61	0.001144	0.1512	813.3	2590.0	814.8	2786.5	2.2512	6.4936
14.00	195.05	0.001149	0.1408	828.5	2591.8	830.1	2788.9	2.2839	6.4675
15.00	198.30	0.001154	0.1317	843.0	2593.5	844.7	2791.0	2.3147	6.4431
16.00	201.38	0.001159	0.1237	856.8	2594.9	858.6	2792.9	2.3438	6.4200
17.00	204.31	0.001163	0.1167	869.9	2596.2	871.9	2794.5	2.3715	6.3983
18.00	207.12	0.001168	0.1104	882.5	2597.3	884.6	2796.0	2.3978	6.3776
19.00	209.81	0.001172	0.1047	894.6	2598.3	896.8	2797.3	2.4229	6.3579
20.00	212.38	0.001177	0.09958	906.3	2599.2	908.6	2798.4	2.4470	6.3392
25.00	223.96	0.001197	0.07995	959.0	2602.2	962.0	2802.0	2.5544	6.2560
30.00	233.86	0.001217	0.06666	1004.7	2603.3	1008.4	2803.3	2.6456	6.1858
35.00	242.56	0.001235	0.05706	1045.5	2603.0	1049.8	2802.7	2.7254	6.1245
40.00	250.36	0.001253	0.04978	1082.4	2601.8	1087.4	2800.9	2.7967	6.0697
45.00	257.44	0.001270	0.04406	1116.4	2599.7	1122.1	2798.0	2.8613	6.0198
50.00	263.94	0.001286	0.03945	1148.1	2597.0	1154.5	2794.2	2.9207	5.9737
55.00	269.97	0.001303	0.03564	1177.8	2593.7	1184.9	2789.7	2.9759	5.9307

P, bar	T ^{0}C	v_f, m^3/kg	v_g, m^3/kg	u_f, kJ/kg	u_g, kJ/kg	h_f, kJ/kg	h_g, kJ/kg	s_f, kJ/kg K	s_g, kJ/kg K
60.00	275.59	0.001319	0.03245	1205.8	2589.9	1213.7	2784.6	3.0274	5.8901
65.00	280.86	0.001336	0.02973	1232.5	2585.6	1241.2	2778.8	3.0760	5.8515
70.00	285.83	0.001352	0.02738	1258.0	2580.9	1267.4	2772.6	3.1220	5.8146
75.00	290.54	0.001368	0.02533	1282.4	2575.8	1292.7	2765.8	3.1658	5.7792
80.00	295.01	0.001385	0.02353	1306.0	2570.4	1317.1	2758.6	3.2077	5.7448
85.00	299.27	0.001401	0.02193	1328.8	2564.6	1340.7	2751.0	3.2478	5.7115
90.00	303.35	0.001418	0.02049	1350.9	2558.4	1363.7	2742.9	3.2866	5.6790
95.00	307.25	0.001435	0.01920	1372.4	2552.0	1386.0	2734.4	3.3240	5.6472
100.00	311.00	0.001453	0.01803	1393.3	2545.1	1407.9	2725.5	3.3603	5.6159
110.00	318.08	0.001489	0.01599	1433.9	2530.5	1450.3	2706.4	3.4300	5.5545
120.00	324.68	0.001526	0.01427	1473.0	2514.4	1491.3	2685.6	3.4965	5.4941
130.00	330.86	0.001566	0.01279	1511.0	2496.7	1531.4	2662.9	3.5606	5.4339
140.00	336.67	0.001610	0.01149	1548.3	2477.2	1570.9	2638.1	3.6230	5.3730
150.00	342.16	0.001657	0.01034	1585.3	2455.8	1610.2	2610.9	3.6844	5.3108
160.00	347.36	0.001710	0.009308	1622.3	2431.9	1649.7	2580.8	3.7457	5.2463
170.00	352.29	0.001769	0.008369	1660.0	2405.1	1690.0	2547.4	3.8077	5.1785
180.00	356.99	0.001839	0.007499	1698.9	2374.6	1732.0	2509.5	3.8717	5.1055
190.00	361.47	0.001925	0.006673	1740.3	2338.6	1776.9	2465.4	3.9396	5.0246
200.00	365.75	0.002039	0.005858	1786.3	2294.2	1827.1	2411.4	4.0154	4.9299
210.00	369.83	0.002212	0.004988	1842.9	2232.8	1889.4	2337.5	4.1093	4.8062
220.00	373.71	0.002750	0.003577	1961.4	2085.5	2021.9	2164.2	4.3109	4.5308

Liquid and Vapor regions for Steam and Water

In the saturated state, pressure and temperature are dependent on each other. When one of these properties is defined the other gets fixed automatically. The P-T saturation diagram illustrates this relationship. The area to the left of the saturation curve defines liquid states whereas the area to the right identifies superheated steam. Above the critical pressure shown as a horizontal line on the following figure, steam is supercritical. Precise liquid and vapor states cannot be defined for supercritical steam since a distinct phase boundary does not exist However, a pseudocritical temperature[3] can be identified that demarcates liquid-like and vapor-like states. The figure below helps in ascertaining the condition of steam if the pressure and temperature is known.

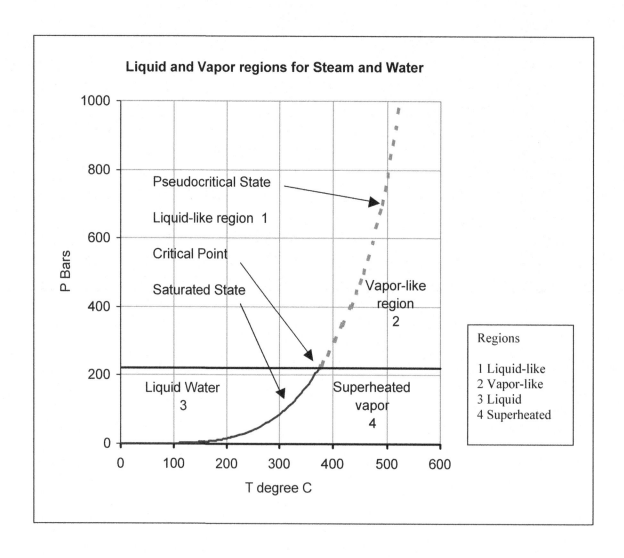

---

[3] *Thermodynamic Properties of Supercritical Steam* by Ashok Malhotra, ISBN: 978-1-4116-8491-1 (or 1-4116-8491-5)

Superheated Steam

	P	0.1	bar,	Ts	45.81	degree C		P	0.5	bar,	Ts	81.32	degree C
T	v	u		h	s		T	v	u		h	s	
50	14.867	2443.3		2592.0	8.1741		90	3.323	2496.4		2662.6	7.6414	
100	17.197	2515.5		2687.4	8.4488		100	3.419	2511.5		2682.4	7.6952	
150	19.514	2587.9		2783.0	8.6892		150	3.890	2585.7		2780.2	7.9412	
200	21.826	2661.3		2879.6	8.9048		200	4.356	2660.0		2877.8	8.1591	
250	24.136	2736.1		2977.4	9.1014		250	4.821	2735.1		2976.2	8.3568	
300	26.446	2812.3		3076.7	9.2827		300	5.284	2811.6		3075.8	8.5386	
350	28.755	2890.0		3177.5	9.4513		350	5.747	2889.4		3176.8	8.7076	
400	31.064	2969.3		3279.9	9.6093		400	6.209	2968.8		3279.3	8.8658	
450	33.372	3050.2		3384.0	9.7584		450	6.672	3049.9		3383.5	9.0150	
500	35.680	3132.9		3489.7	9.8997		500	7.134	3132.6		3489.2	9.1565	
550	37.988	3217.2		3597.1	10.0343		550	7.596	3216.9		3596.7	9.2912	
600	40.296	3303.3		3706.3	10.1631		600	8.058	3303.1		3706.0	9.4200	
650	42.604	3391.2		3817.2	10.2866		650	8.520	3391.0		3816.9	9.5436	
700	44.912	3480.8		3929.9	10.4055		700	8.981	3480.6		3929.7	9.6625	
750	47.220	3572.2		4044.4	10.5202		750	9.443	3572.0		4044.2	9.7772	
800	49.528	3665.3		4160.6	10.6311		800	9.905	3665.2		4160.4	9.8882	
	P	1	bar,	Ts	99.61	degree C		P	2	bar,	Ts	120.21	degree C
T	v	u		h	s		T	v	u		h	s	
100	1.696	2506.2		2675.8	7.3610		125	0.898	2537.0		2716.6	7.1530	
150	1.937	2582.9		2776.6	7.6147		150	0.960	2577.1		2769.1	7.2809	
200	2.172	2658.2		2875.5	7.8356		200	1.081	2654.7		2870.8	7.5081	
250	2.406	2733.9		2974.5	8.0346		250	1.199	2731.5		2971.3	7.7100	
300	2.639	2810.7		3074.5	8.2171		300	1.316	2808.8		3072.1	7.8940	
350	2.871	2888.7		3175.8	8.3865		350	1.433	2887.3		3173.9	8.0643	
400	3.103	2968.3		3278.5	8.5451		400	1.549	2967.1		3277.0	8.2235	
450	3.334	3049.4		3382.8	8.6945		450	1.665	3048.4		3381.5	8.3733	
500	3.566	3132.2		3488.7	8.8361		500	1.781	3131.3		3487.6	8.5151	
550	3.797	3216.6		3596.3	8.9709		550	1.897	3215.9		3595.4	8.6501	
600	4.028	3302.8		3705.6	9.0998		600	2.013	3302.2		3704.8	8.7792	
650	4.259	3390.7		3816.6	9.2234		650	2.129	3390.2		3815.9	8.9029	
700	4.490	3480.4		3929.4	9.3424		700	2.244	3479.9		3928.8	9.0220	
750	4.721	3571.8		4043.9	9.4571		750	2.360	3571.4		4043.4	9.1368	
800	4.952	3665.0		4160.2	9.5681		800	2.476	3664.7		4159.8	9.2479	

	P	3	bar,	Ts	133.53	degree C			P	4	bar,	Ts	143.61	degree C
T		v	u	h		s		T		v	u	h		s
135		0.608	2545.7	2728.2		6.9997		145		0.464	2555.6	2741.3		6.9032
150		0.634	2571.0	2761.2		7.0791		150		0.471	2564.4	2752.8		6.9305
200		0.716	2651.0	2866.0		7.3132		200		0.534	2647.3	2861.0		7.1724
250		0.796	2729.0	2967.9		7.5181		250		0.595	2726.5	2964.6		7.3805
300		0.875	2807.0	3069.6		7.7037		300		0.655	2805.2	3067.1		7.5677
350		0.954	2885.9	3172.0		7.8749		350		0.714	2884.4	3170.0		7.7398
400		1.032	2966.0	3275.4		8.0346		400		0.773	2964.8	3273.9		7.9001
450		1.109	3047.5	3380.2		8.1848		450		0.831	3046.5	3379.0		8.0507
500		1.187	3130.5	3486.6		8.3269		500		0.889	3129.7	3485.5		8.1931
550		1.264	3215.2	3594.5		8.4622		550		0.948	3214.5	3593.6		8.3286
600		1.341	3301.6	3704.0		8.5914		600		1.006	3301.0	3703.2		8.4579
650		1.419	3389.7	3815.3		8.7152		650		1.064	3389.2	3814.6		8.5819
700		1.496	3479.5	3928.2		8.8344		700		1.122	3479.0	3927.6		8.7012
750		1.573	3571.0	4042.9		8.9493		750		1.179	3570.6	4042.4		8.8161
800		1.650	3664.3	4159.3		9.0604		800		1.237	3663.9	4158.9		8.9273
	P	5	bar,	Ts	151.84	degree C			P	6	bar,	Ts	158.83	degree C
T		v	u	h		s		T		v	u	h		s
155		0.3783	2566.5	2755.7		6.8383		160		0.3167	2569.0	2759.0		6.7658
200		0.4250	2643.4	2855.9		7.0611		200		0.3521	2639.4	2850.7		6.9684
250		0.4744	2723.9	2961.1		7.2726		250		0.3939	2721.3	2957.7		7.1834
300		0.5226	2803.3	3064.6		7.4614		300		0.4344	2801.4	3062.1		7.3740
350		0.5701	2883.0	3168.1		7.6345		350		0.4743	2881.5	3166.1		7.5480
400		0.6173	2963.6	3272.3		7.7954		400		0.5137	2962.5	3270.7		7.7095
450		0.6642	3045.6	3377.7		7.9464		450		0.5530	3044.6	3376.4		7.8609
500		0.7109	3128.9	3484.4		8.0891		500		0.5920	3128.1	3483.3		8.0039
550		0.7576	3213.9	3592.6		8.2247		550		0.6309	3213.2	3591.7		8.1398
600		0.8041	3300.4	3702.5		8.3543		600		0.6698	3299.8	3701.7		8.2694
650		0.8506	3388.6	3813.9		8.4784		650		0.7085	3388.1	3813.2		8.3937
700		0.8970	3478.6	3927.0		8.5977		700		0.7473	3478.1	3926.5		8.5131
750		0.9433	3570.2	4041.9		8.7128		750		0.7859	3569.8	4041.4		8.6282
800		0.9897	3663.6	4158.4		8.8240		800		0.8246	3663.2	4157.9		8.7395
	P	7	bar,	Ts	164.95	degree C			P	8	bar,	Ts	170.41	degree C
T		v	u	h		s		T		v	u	h		s
165		0.2728	2571.9073	2762.9		6.7073		175		0.2436	2585.1	2780.0		6.6878
200		0.3000	2635.2938	2845.3		6.8884		200		0.2609	2631.1	2839.8		6.8176
250		0.3364	2718.6692	2954.1		7.1071		250		0.2932	2716.0	2950.5		7.0403
300		0.3714	2799.5173	3059.5		7.2995		300		0.3242	2797.6	3056.9		7.2345
350		0.4058	2880.0858	3164.1		7.4745		350		0.3544	2878.6	3162.2		7.4106
400		0.4398	2961.3135	3269.1		7.6366		400		0.3843	2960.1	3267.6		7.5733
450		0.4735	3043.6455	3375.1		7.7884		450		0.4139	3042.7	3373.8		7.7255
500		0.5070	3127.3213	3482.3		7.9317		500		0.4433	3126.5	3481.2		7.8690
550		0.5405	3212.4843	3590.8		8.0678		550		0.4726	3211.8	3589.9		8.0053

600	0.5738	3299.2257	3700.9	8.1976	600	0.5019	3298.6	3700.1	8.1353
650	0.6071	3387.6040	3812.6	8.3220	650	0.5310	3387.1	3811.9	8.2598
700	0.6403	3477.6549	3925.9	8.4415	700	0.5601	3477.2	3925.3	8.3794
750	0.6735	3569.3980	4040.8	8.5567	750	0.5892	3569.0	4040.3	8.4947
800	0.7066	3662.8447	4157.5	8.6680	800	0.6182	3662.5	4157.0	8.6060

	P	9	bar,	Ts	175.36	degree C	P	10	bar,	Ts	179.89	degree C
T	v	u	h	s			T	v	u	h	s	
180	0.2179	2589.0	2785.2	6.6481			190	0.2003	2603.2	2803.5	6.6426
200	0.2304	2626.7	2834.1	6.7538			200	0.2060	2622.3	2828.3	6.6955
250	0.2596	2713.3	2946.9	6.9806			250	0.2327	2710.5	2943.2	6.9266
300	0.2874	2795.7	3054.3	7.1768			300	0.2580	2793.7	3051.7	7.1247
350	0.3145	2877.2	3160.2	7.3538			350	0.2825	2875.7	3158.2	7.3028
400	0.3411	2959.0	3266.0	7.5172			400	0.3066	2957.8	3264.4	7.4668
450	0.3675	3041.7	3372.5	7.6698			450	0.3304	3040.8	3371.2	7.6198
500	0.3938	3125.7	3480.1	7.8136			500	0.3541	3124.9	3479.0	7.7640
550	0.4199	3211.1	3589.0	7.9501			550	0.3777	3210.4	3588.1	7.9007
600	0.4459	3298.0	3699.3	8.0803			600	0.4011	3297.4	3698.6	8.0309
650	0.4718	3386.6	3811.2	8.2049			650	0.4245	3386.1	3810.5	8.1557
700	0.4977	3476.7	3924.7	8.3246			700	0.4478	3476.3	3924.1	8.2755
750	0.5236	3568.6	4039.8	8.4399			750	0.4711	3568.2	4039.3	8.3909
800	0.5494	3662.1	4156.6	8.5513			800	0.4944	3661.8	4156.1	8.5024

	P	12	bar,	Ts	187.96	degree C	P	14	bar,	Ts	195.05	degree C
T	v	u	h	s			T	v	u	h	s	
190	0.1643	2592.3	2789.4	6.5340			200	0.1430	2602.8	2803.0	6.4975
200	0.1693	2612.9	2816.1	6.5908			225	0.1536	2652.9	2867.9	6.6313
250	0.1924	2704.8	2935.7	6.8314			250	0.1635	2699.0	2927.9	6.7488
300	0.2139	2789.8	3046.4	7.0336			300	0.1823	2785.8	3041.0	6.9553
350	0.2345	2872.7	3154.1	7.2138			350	0.2003	2869.7	3150.1	7.1378
400	0.2548	2955.4	3261.2	7.3791			400	0.2178	2953.0	3258.0	7.3044
450	0.2748	3038.8	3368.6	7.5330			450	0.2351	3036.9	3366.0	7.4591
500	0.2946	3123.3	3476.8	7.6777			500	0.2522	3121.6	3474.7	7.6045
550	0.3143	3209.0	3586.2	7.8148			550	0.2691	3207.7	3584.4	7.7420
600	0.3339	3296.3	3697.0	7.9454			600	0.2860	3295.1	3695.4	7.8729
650	0.3535	3385.0	3809.2	8.0704			650	0.3028	3384.0	3807.8	7.9981
700	0.3730	3475.4	3922.9	8.1904			700	0.3195	3474.5	3921.8	8.1183
750	0.3924	3567.4	4038.3	8.3059			750	0.3362	3566.6	4037.2	8.2340
800	0.4118	3661.0	4155.2	8.4175			800	0.3529	3660.3	4154.3	8.3457

	P	16	bar,	Ts	201.38	degree C	P	18	bar,	Ts	207.12	degree C
T	v	u	h	s			T	v	u	h	s	
210	0.1272	2614.2	2817.7	6.4720			210	0.1114	2604.2	2804.8	6.3958
211	0.1276	2616.4	2820.5	6.4777			211	0.1118	2606.5	2807.8	6.4020
212	0.1280	2618.5	2823.3	6.4834			212	0.1122	2608.8	2810.7	6.4081
213	0.1284	2620.6	2826.0	6.4890			213	0.1125	2611.1	2813.6	6.4141
214	0.1288	2622.7	2828.7	6.4946			214	0.1129	2613.3	2816.5	6.4201

T	v	u	h	s		T	v	u	h	s
215	0.1291	2624.8	2831.4	6.5002		215	0.1133	2615.6	2819.4	6.4260
216	0.1295	2626.9	2834.1	6.5057		216	0.1136	2617.8	2822.3	6.4319
217	0.1299	2629.0	2836.8	6.5111		217	0.1140	2620.0	2825.1	6.4377
218	0.1303	2631.0	2839.5	6.5166		218	0.1143	2622.2	2827.9	6.4434
219	0.1307	2633.1	2842.1	6.5220		219	0.1147	2624.3	2830.8	6.4491
220	0.1310	2635.1	2844.8	6.5273		220	0.1150	2626.5	2833.5	6.4548
221	0.1314	2637.1	2847.4	6.5327		221	0.1154	2628.6	2836.3	6.4604
222	0.1318	2639.1	2850.0	6.5380		222	0.1157	2630.7	2839.1	6.4659

P	20	bar,	Ts	212.38	degree C	P	30	bar,	Ts	233.86	degree C
T	v	u	h	s		T	v	u	h	s	
220	0.1022	2617.3	2821.7	6.3868		240	0.0682	2619.9	2824.6	6.2275
250	0.1115	2680.3	2903.2	6.5474		250	0.0706	2644.7	2856.5	6.2893
300	0.1255	2773.2	3024.3	6.7685		300	0.0812	2750.8	2994.3	6.5412
350	0.1386	2860.5	3137.6	6.9582		350	0.0906	2844.4	3116.1	6.7449
400	0.1512	2945.8	3248.2	7.1290		400	0.0994	2933.4	3231.6	6.9233
450	0.1635	3031.0	3358.1	7.2863		450	0.1079	3021.0	3344.7	7.0853
500	0.1757	3116.7	3468.1	7.4335		500	0.1162	3108.5	3457.0	7.2356
550	0.1877	3203.5	3578.9	7.5723		550	0.1244	3196.5	3569.6	7.3767
600	0.1996	3291.5	3690.7	7.7042		600	0.1324	3285.5	3682.8	7.5102
650	0.2115	3380.9	3803.8	7.8301		650	0.1405	3375.6	3797.0	7.6373
700	0.2233	3471.7	3918.2	7.9509		700	0.1484	3467.1	3912.3	7.7590
750	0.2350	3564.1	4034.2	8.0670		750	0.1563	3560.1	4029.0	7.8759
800	0.2467	3658.1	4151.6	8.1791		800	0.1642	3654.5	4147.0	7.9885

P	40	bar,	Ts	250.36	degree C	P	50	bar,	Ts	263.94	degree C
T	v	u	h	s		T	v	u	h	s	
260	0.0518	2630.1	2837.2	6.1384		270	0.04057	2617.0	2819.8	6.0211
300	0.0589	2726.2	2961.7	6.3638		300	0.04535	2698.9	2925.6	6.2109
350	0.0665	2827.4	3093.3	6.5843		350	0.05197	2809.4	3069.3	6.4515
400	0.0734	2920.6	3214.4	6.7712		400	0.05784	2907.4	3196.6	6.6481
450	0.0800	3010.8	3331.0	6.9383		450	0.06332	3000.4	3317.0	6.8208
500	0.0864	3100.1	3445.8	7.0919		500	0.06858	3091.6	3434.5	6.9778
550	0.0927	3189.4	3560.2	7.2353		550	0.07369	3182.3	3550.8	7.1235
600	0.0989	3279.4	3674.8	7.3704		600	0.07870	3273.3	3666.8	7.2604
650	0.1049	3370.4	3790.2	7.4989		650	0.08364	3365.1	3783.3	7.3901
700	0.1110	3462.5	3906.4	7.6215		700	0.08851	3457.9	3900.5	7.5137
750	0.1170	3556.0	4023.8	7.7391		750	0.09335	3551.8	4018.6	7.6321
800	0.1229	3650.8	4142.5	7.8523		800	0.09815	3647.1	4137.9	7.7459

P	60	bar,	Ts	275.59	degree C	P	70	bar,	Ts	285.83	degree C
T	v	u	h	s		T	v	u	h	s	
280	0.03320	2606.0	2805.2	5.9276		290	0.02804	2597.7	2794.0	5.8528
300	0.03619	2668.3	2885.5	6.0702		300	0.02949	2633.4	2839.8	5.9335
350	0.04225	2790.3	3043.9	6.3356		350	0.03527	2770.0	3016.8	6.2303
400	0.04742	2893.6	3178.2	6.5431		400	0.03996	2879.4	3159.1	6.4501
450	0.05217	2989.8	3302.8	6.7216		450	0.04419	2978.8	3288.2	6.6351

T	v	u	h	s		T	v	u	h	s
500	0.05667	3082.9	3422.9	6.8824		500	0.04816	3074.1	3411.3	6.7997
550	0.06102	3175.1	3541.2	7.0306		550	0.05197	3167.8	3531.5	6.9505
600	0.06526	3267.2	3658.8	7.1692		600	0.05566	3261.0	3650.6	7.0909
650	0.06943	3359.8	3776.4	7.3002		650	0.05928	3354.4	3769.4	7.2232
700	0.07354	3453.2	3894.5	7.4248		700	0.06285	3448.5	3888.5	7.3488
750	0.07761	3547.7	4013.4	7.5439		750	0.06637	3543.6	4008.1	7.4687
800	0.08164	3643.4	4133.3	7.6583		800	0.06985	3639.7	4128.7	7.5837

P	80	bar,	Ts	295.01	degree C		P	90	bar,	Ts	303.35	degree C
T	v	u	h	s			T	v	u	h	s	
300	0.02428	2592.1	2786.4	5.7935		310	0.02145	2589.6	2782.6	5.7475
350	0.02998	2748.2	2988.1	6.1319		350	0.02582	2724.9	2957.2	6.0378
400	0.03435	2864.5	3139.3	6.3657		400	0.02996	2849.1	3118.8	6.2875
450	0.03820	2967.7	3273.2	6.5577		450	0.03353	2956.2	3257.9	6.4871
500	0.04177	3065.2	3399.4	6.7264		500	0.03680	3056.2	3387.3	6.6601
550	0.04517	3160.4	3521.8	6.8798		550	0.03989	3152.9	3511.9	6.8163
600	0.04846	3254.7	3642.4	7.0221		600	0.04286	3248.4	3634.2	6.9605
650	0.05167	3349.0	3762.4	7.1557		650	0.04575	3343.6	3755.4	7.0955
700	0.05483	3443.8	3882.4	7.2823		700	0.04859	3439.1	3876.4	7.2231
750	0.05793	3539.4	4002.9	7.4030		750	0.05137	3535.2	3997.6	7.3446
800	0.06101	3636.0	4124.0	7.5186		800	0.05413	3632.2	4119.4	7.4608

P	100	bar,	Ts	311.00	degree C		P	110	bar,	Ts	318.08	degree C
T	v	u	h	s			T	v	u	h	s	
320	0.01927	2589.9	2782.7	5.7131		320	0.01628	2542.0	2721.1	5.5793
350	0.02244	2699.5	2924.0	5.9458		350	0.01963	2671.9	2887.8	5.8541
400	0.02644	2833.0	3097.4	6.2139		400	0.02354	2816.2	3075.1	6.1438
450	0.02978	2944.4	3242.3	6.4217		450	0.02672	2932.4	3226.2	6.3605
500	0.03281	3046.9	3375.1	6.5993		500	0.02955	3037.6	3362.6	6.5430
550	0.03566	3145.4	3501.9	6.7584		550	0.03219	3137.8	3491.9	6.7050
600	0.03838	3242.1	3625.8	6.9045		600	0.03471	3235.7	3617.5	6.8531
650	0.04102	3338.2	3748.3	7.0409		650	0.03714	3332.7	3741.2	6.9910
700	0.04359	3434.3	3870.3	7.1696		700	0.03951	3429.5	3864.2	7.1207
750	0.04613	3531.0	3992.3	7.2918		750	0.04183	3526.8	3987.0	7.2437
800	0.04862	3628.5	4114.7	7.4087		800	0.04412	3624.7	4110.1	7.3612

P	120	bar,	Ts	324.68	degree C		P	130	bar,	Ts	330.86	degree C
T	v	u	h	s			T	v	u	h	s	
330	0.01502	2547.9	2728.1	5.5650		340	0.01403	2556.5	2738.9	5.5589
350	0.01722	2641.3	2848.0	5.7607		350	0.01512	2607.1	2803.6	5.6635
400	0.02111	2798.6	3051.9	6.0762		400	0.01903	2780.2	3027.6	6.0104
450	0.02415	2920.0	3209.8	6.3027		450	0.02198	2907.2	3192.9	6.2475
500	0.02683	3028.0	3350.0	6.4902		500	0.02452	3018.3	3337.1	6.4404
550	0.02930	3130.0	3481.7	6.6553		550	0.02686	3122.2	3471.4	6.6087
600	0.03165	3229.2	3609.0	6.8055		600	0.02906	3222.7	3600.5	6.7610
650	0.03391	3327.1	3734.1	6.9448		650	0.03118	3321.6	3726.9	6.9018
700	0.03611	3424.7	3858.0	7.0756		700	0.03323	3419.9	3851.9	7.0336

(continuation of previous tables)

T	v	u	h	s		T	v	u	h	s
750	0.03826	3522.6	3981.6	7.1994		750	0.03523	3518.3	3976.3	7.1583
800	0.04037	3621.0	4105.4	7.3175		800	0.03720	3617.2	4100.7	7.2771

	P 140 bar, Ts 336.67 degree C						P 150 bar, Ts 342.16 degree C			
T	v	u	h	s		T	v	u	h	s
340	0.01200	2504.4	2672.4	5.4291		345	0.01080	2482.5	2644.5	5.3653
350	0.01323	2567.7	2752.9	5.5595		350	0.01148	2520.8	2693.0	5.4435
400	0.01724	2760.9	3002.2	5.9457		400	0.01567	2740.5	2975.5	5.8817
450	0.02010	2894.1	3175.6	6.1945		450	0.01848	2880.7	3157.8	6.1433
500	0.02255	3008.4	3324.1	6.3931		500	0.02083	2998.4	3310.8	6.3479
550	0.02476	3114.3	3461.0	6.5648		550	0.02295	3106.3	3450.5	6.5230
600	0.02684	3216.1	3591.9	6.7192		600	0.02492	3209.5	3583.3	6.6797
650	0.02883	3316.0	3719.7	6.8615		650	0.02680	3310.4	3712.4	6.8235
700	0.03076	3415.1	3845.7	6.9944		700	0.02862	3410.2	3839.5	6.9576
750	0.03264	3514.0	3970.9	7.1200		750	0.03039	3509.8	3965.6	7.0839
800	0.03448	3613.4	4096.0	7.2393		800	0.03212	3609.6	4091.3	7.2039

	P 160 bar, Ts 347.36 degree C						P 180 bar, Ts 356.99 degree C			
T	v	u	h	s		T	v	u	h	s
350	0.00977	2460.7	2617.0	5.3045		360	0.00811	2420.1	2566.0	5.1950
400	0.01428	2719.0	2947.5	5.8177		400	0.01191	2671.8	2886.3	5.6881
450	0.01705	2866.8	3139.6	6.0935		450	0.01465	2837.9	3101.7	5.9973
500	0.01932	2988.1	3297.3	6.3045		500	0.01681	2967.1	3269.7	6.2222
550	0.02135	3098.2	3439.8	6.4832		550	0.01870	3081.7	3418.3	6.4085
600	0.02324	3202.8	3574.6	6.6422		600	0.02043	3189.3	3557.0	6.5722
650	0.02503	3304.7	3705.1	6.7876		650	0.02206	3293.3	3690.4	6.7208
700	0.02675	3405.3	3833.3	6.9228		700	0.02363	3395.4	3820.7	6.8583
750	0.02842	3505.5	3960.2	7.0499		750	0.02514	3496.8	3949.4	6.9872
800	0.03006	3605.7	4086.6	7.1706		800	0.02662	3598.1	4077.2	7.1091

	P 200 bar, Ts 365.75 degree C						P 220 bar, Ts 373.71 degree C			
T	v	u	h	s		T	v	u	h	s
370	0.00692	2388.0	2526.5	5.1095		380	0.00612	2369.8	2504.6	5.0556
400	0.00995	2617.8	2816.8	5.5525		400	0.00826	2554.2	2735.8	5.4050
450	0.01272	2807.1	3061.5	5.9041		450	0.01112	2774.4	3019.0	5.8124
500	0.01479	2945.3	3241.2	6.1445		500	0.01314	2922.8	3211.8	6.0704
550	0.01657	3064.8	3396.2	6.3390		550	0.01483	3047.6	3373.8	6.2736
600	0.01818	3175.5	3539.2	6.5077		600	0.01635	3161.6	3521.2	6.4475
650	0.01969	3281.7	3675.6	6.6596		650	0.01776	3270.0	3660.6	6.6029
700	0.02113	3385.5	3808.2	6.7994		700	0.01909	3375.5	3795.5	6.7451
750	0.02252	3488.1	3938.5	6.9301		750	0.02038	3479.4	3927.6	6.8776
800	0.02387	3590.4	4067.7	7.0534		800	0.02162	3582.6	4058.2	7.0022

Supercritical Steam

P	225	bar			P	230	bar		
T	v	u	h	s	T	v	u	h	s
275	0.00128	1177.3	1206.1	2.9745	275	0.00128	1176.7	1206.1	2.9732
300	0.00135	1302.1	1332.5	3.1999	300	0.00135	1301.1	1332.2	3.1982
325	0.00146	1438.0	1470.8	3.4360	325	0.00145	1436.6	1470.0	3.4335
350	0.00163	1597.0	1633.7	3.7025	350	0.00162	1594.2	1631.6	3.6978
375	0.00247	1915.9	1971.5	4.2310	375	0.00222	1862.8	1913.8	4.1402
400	0.00786	2536.2	2713.1	5.3654	400	0.00748	2517.2	2689.2	5.3242
425	0.00951	2669.3	2883.3	5.6139	425	0.00916	2657.6	2868.3	5.5859
450	0.01076	2765.8	3008.0	5.7896	450	0.01042	2757.2	2996.8	5.7668
475	0.01182	2846.1	3112.1	5.9312	475	0.01148	2839.1	3103.1	5.9114
500	0.01277	2917.0	3204.3	6.0524	500	0.01242	2911.2	3196.7	6.0345
525	0.01363	2982.0	3288.8	6.1600	525	0.01327	2977.0	3282.3	6.1434
550	0.01444	3043.2	3368.1	6.2578	550	0.01407	3038.8	3362.4	6.2422
575	0.01521	3101.6	3443.7	6.3483	575	0.01483	3097.7	3438.6	6.3335
600	0.01594	3158.0	3516.6	6.4331	600	0.01555	3154.5	3512.1	6.4188
625	0.01664	3213.1	3587.5	6.5131	625	0.01624	3209.9	3583.4	6.4994
650	0.01732	3267.1	3656.9	6.5893	650	0.01691	3264.1	3653.1	6.5759
675	0.01799	3320.3	3725.0	6.6622	675	0.01757	3317.6	3721.6	6.6491
700	0.01864	3373.0	3792.3	6.7322	700	0.01820	3370.4	3789.1	6.7195
725	0.01927	3425.2	3858.9	6.7997	725	0.01883	3422.9	3855.9	6.7872
750	0.01990	3477.2	3924.9	6.8651	750	0.01944	3475.0	3922.2	6.8528
775	0.02051	3529.0	3990.5	6.9284	775	0.02005	3526.9	3988.0	6.9163
800	0.02112	3580.7	4055.9	6.9900	800	0.02064	3578.7	4053.5	6.9781
P	240	bar			P	250	bar		
T	v	u	h	s	T	v	u	h	s
275	0.00128	1175.3	1205.9	2.9707	275	0.00127	1174.0	1205.8	2.9681
300	0.00135	1299.2	1331.6	3.1948	300	0.00135	1297.4	1331.1	3.1915
325	0.00145	1433.8	1468.5	3.4286	325	0.00144	1431.0	1467.1	3.4238
350	0.00161	1588.9	1627.6	3.6888	350	0.00160	1583.9	1623.9	3.6803
375	0.00206	1823.0	1872.5	4.0731	375	0.00198	1799.9	1849.3	4.0344
400	0.00673	2475.8	2637.4	5.2366	400	0.00600	2428.5	2578.6	5.1399
425	0.00850	2633.4	2837.4	5.5289	425	0.00789	2607.8	2804.9	5.4706
450	0.00977	2739.4	2974.0	5.7212	450	0.00918	2721.0	2950.4	5.6755
475	0.01083	2825.0	3084.8	5.8720	475	0.01023	2810.5	3066.1	5.8330
500	0.01175	2899.4	3181.4	5.9991	500	0.01114	2887.4	3165.9	5.9642
525	0.01259	2966.9	3269.1	6.1107	525	0.01197	2956.6	3255.9	6.0787
550	0.01338	3029.9	3350.9	6.2116	550	0.01274	3020.9	3339.3	6.1816
575	0.01411	3089.7	3428.4	6.3044	575	0.01346	3081.8	3418.1	6.2760

T	v	u	h	s		T	v	u	h	s
600	0.01481	3147.4	3502.9	6.3910		600	0.01414	3140.2	3493.7	6.3638
625	0.01549	3203.4	3575.1	6.4725		625	0.01480	3196.9	3566.8	6.4464
650	0.01614	3258.2	3645.6	6.5499		650	0.01543	3252.2	3638.0	6.5246
675	0.01677	3312.1	3714.7	6.6237		675	0.01604	3306.6	3707.7	6.5991
700	0.01739	3365.4	3782.8	6.6946		700	0.01664	3360.3	3776.4	6.6706
725	0.01800	3418.1	3850.0	6.7629		725	0.01723	3413.4	3844.1	6.7393
750	0.01859	3470.6	3916.7	6.8289		750	0.01780	3466.2	3911.2	6.8057
775	0.01917	3522.8	3982.9	6.8928		775	0.01837	3518.6	3977.8	6.8700
800	0.01975	3574.8	4048.8	6.9549		800	0.01892	3570.9	4044.0	6.9324

P	300	bar				P	350	bar		
T	v	u	h	s		T	v	u	h	s
300	0.00133	1288.7	1328.7	3.1756		300	0.00132	1280.6	1326.8	3.1608
350	0.00155	1562.2	1608.8	3.6435		350	0.00152	1544.4	1597.5	3.6131
400	0.00280	2068.5	2152.4	4.4750		400	0.00211	1914.7	1988.4	4.2140
450	0.00674	2618.8	2820.9	5.4419		450	0.00496	2497.4	2671.0	5.1945
500	0.00869	2824.1	3084.8	5.7956		500	0.00693	2755.3	2998.0	5.6331
550	0.01017	2974.6	3279.8	6.0403		550	0.00835	2925.9	3218.1	5.9093
600	0.01144	3103.5	3446.9	6.2374		600	0.00952	3065.7	3399.0	6.1229
650	0.01259	3222.0	3599.7	6.4077		650	0.01057	3191.1	3560.9	6.3032
700	0.01365	3334.6	3744.2	6.5602		700	0.01152	3308.6	3711.9	6.4625
750	0.01466	3443.9	3883.8	6.7000		750	0.01242	3421.5	3856.3	6.6072
800	0.01563	3551.4	4020.2	6.8303		800	0.01328	3531.7	3996.5	6.7411

P	400	bar				P	450	bar		
T	v	u	h	s		T	v	u	h	s
300	0.00131	1273.1	1325.4	3.1469		300	0.00130	1266.0	1324.4	3.1338
350	0.00149	1529.2	1588.7	3.5870		350	0.00146	1515.8	1581.7	3.5638
400	0.00191	1854.7	1931.1	4.1141		400	0.00180	1816.4	1897.6	4.0505
450	0.00369	2364.1	2511.8	4.9447		450	0.00292	2246.1	2377.3	4.7362
500	0.00562	2681.7	2906.7	5.4746		500	0.00463	2604.8	2813.4	5.3209
550	0.00699	2875.2	3154.6	5.7859		550	0.00594	2823.0	3090.2	5.6685
600	0.00809	3026.9	3350.4	6.0170		600	0.00698	2987.3	3301.5	5.9179
650	0.00905	3159.6	3521.8	6.2079		650	0.00788	3127.7	3482.5	6.1197
700	0.00993	3282.2	3679.4	6.3743		700	0.00870	3255.6	3647.0	6.2932
750	0.01075	3398.8	3828.8	6.5239		750	0.00945	3376.1	3801.3	6.4479
800	0.01152	3511.9	3972.8	6.6614		800	0.01016	3492.1	3949.3	6.5891

P	500	bar				P	1000	bar		
T	v	u	h	s		T	v	u	h	s
300	0.00129	1259.3	1323.7	3.1214		300	0.00121	1207.4	1328.9	3.0215
350	0.00144	1503.9	1576.0	3.5430		350	0.00131	1422.7	1553.9	3.3978
400	0.00173	1787.8	1874.3	4.0028		400	0.00144	1646.8	1791.1	3.7638
450	0.00249	2160.1	2284.4	4.5892		450	0.00163	1881.7	2044.5	4.1267
500	0.00389	2528.0	2722.5	5.1759		500	0.00189	2126.9	2316.2	4.4899
550	0.00512	2769.8	3025.7	5.5566		550	0.00225	2371.1	2596.1	4.8407
600	0.00611	2947.2	3252.6	5.8245		600	0.00267	2597.8	2865.1	5.1580

650	0.00696	3095.6	3443.5	6.0372		650	0.00311	2799.2	3110.6	5.4316
700	0.00772	3228.9	3614.8	6.2180		700	0.00355	2976.1	3330.8	5.6640
750	0.00842	3353.3	3774.1	6.3777		750	0.00395	3135.4	3530.7	5.8644
800	0.00907	3472.3	3926.0	6.5226		800	0.00434	3281.6	3715.2	6.0405

Compressed Liquid

	P	0.1	bar, Ts	45.81	degree C		P	1	bar, Ts	99.61	degree C
T	v	u	h	s		T	v	u	h	s	
0.01	0.001000	0.0002	0.0102	0.0000		0.01	0.001000	0.0018	0.1019	0.0000	
5	0.001000	21.02	21.03	0.0763		10	0.001000	42.02	42.12	0.1511	
10	0.001000	42.02	42.03	0.1511		20	0.001002	83.91	84.01	0.2965	
15	0.001001	62.98	62.99	0.2245		30	0.001004	125.73	125.83	0.4368	
20	0.001002	83.92	83.93	0.2965		40	0.001008	167.52	167.62	0.5724	
25	0.001003	104.8	104.8	0.3673		50	0.001012	209.31	209.41	0.7038	
30	0.001004	125.7	125.8	0.4368		60	0.001017	251.12	251.22	0.8312	
35	0.001006	146.6	146.6	0.5052		70	0.001023	292.97	293.07	0.9550	
40	0.001008	167.5	167.5	0.5724		80	0.001029	334.89	334.99	1.0754	

	P	5	bar, Ts	151.84	degree C		P	10	bar, Ts	179.89	degree C
T	v	u	h	s		T	v	u	h	s	
0.01	0.001000	0.0092	0.5092	0.0000		0.01	0.001000	0.0183	1.018	0.0001	
20	0.001002	83.89	84.39	0.2964		20	0.001001	83.86	84.86	0.2963	
40	0.001008	167.5	168.0	0.5722		40	0.001007	167.41	168.42	0.5720	
60	0.001017	251.0	251.6	0.8310		60	0.001017	250.96	251.98	0.8307	
80	0.001029	334.8	335.3	1.0751		80	0.001029	334.68	335.71	1.0748	
100	0.001043	418.9	419.4	1.3067		100	0.001043	418.73	419.77	1.3063	
120	0.001060	503.5	504.0	1.5275		120	0.001060	503.29	504.35	1.5271	
140	0.001080	588.8	589.3	1.7391		170	0.001114	718.21	719.32	2.0417	

	P	20	bar, Ts	212.38	degree C		P	30	bar, Ts	233.86	degree C
T	v	u	h	s		T	v	u	h	s	
0.01	0.000999	0.0360	2.034	0.0001		0.01	0.000999	0.0532	3.049	0.0002	
20	0.001001	83.80	85.80	0.2961		20	0.001000	83.74	86.74	0.2959	
40	0.001007	167.3	169.3	0.5717		40	0.001007	167.17	170.19	0.5713	
60	0.001016	250.8	252.8	0.8302		60	0.001016	250.61	253.66	0.8296	
80	0.001028	334.4	336.5	1.0741		80	0.001028	334.22	337.30	1.0734	
100	0.001042	418.4	420.5	1.3055		100	0.001042	418.15	421.28	1.3048	
120	0.001059	502.9	505.1	1.5262		120	0.001059	502.58	505.76	1.5253	
140	0.001079	588.1	590.3	1.7376		140	0.001078	587.68	590.91	1.7365	
160	0.001101	674.2	676.4	1.9411		160	0.001100	673.67	676.97	1.9399	
180	0.001127	761.4	763.7	2.1382		180	0.001126	760.82	764.20	2.1368	
200	0.001156	850.3	852.6	2.3301		200	0.001155	849.51	852.98	2.3285	
210	0.001173	895.4	897.8	2.4246		220	0.001189	940.26	943.83	2.5165	

	P	40	bar, Ts	250.36	degree C		P	50	bar, Ts	263.94	degree C
T	v	u	h	s		T	v	u	h	s	
0.01	0.000998	0.0699	4.0626	0.0002		0.01	0.000998	0.0860	5.0745	0.0003	
20	0.001000	83.7	87.7	0.2957		20	0.001000	83.62	88.61	0.2955	
40	0.001006	167.1	171.1	0.5709		40	0.001006	166.93	171.96	0.5705	

T	v	u	h	s
60	0.001015	250.4	254.5	0.8291
80	0.001027	334.0	338.1	1.0728
100	0.001041	417.9	422.0	1.3040
120	0.001058	502.2	506.5	1.5244
140	0.001077	587.3	591.6	1.7355
160	0.001100	673.2	677.6	1.9388
180	0.001125	760.2	764.7	2.1354
200	0.001154	848.8	853.4	2.3269
220	0.001188	939.3	944.1	2.5147
240	0.001228	1032.7	1037.6	2.7005
250	0.001252	1080.7	1085.7	2.7933

T	v	u	h	s
60	0.001015	250.26	255.33	0.8286
80	0.001027	333.76	338.89	1.0721
100	0.001041	417.58	422.78	1.3032
120	0.001058	501.88	507.17	1.5235
140	0.001077	586.83	592.22	1.7345
160	0.001099	672.65	678.14	1.9376
180	0.001124	759.60	765.22	2.1341
200	0.001153	848.04	853.80	2.3254
220	0.001187	938.45	944.38	2.5129
240	0.001227	1031.5	1037.7	2.6983
260	0.001275	1128.4	1134.8	2.8839

P 60 bar, Ts 275.59 degree C				
T	v	u	h	s
0.01	0.000997	0.1017	6.0848	0.0003
25	0.001000	104.38	110.38	0.3657
50	0.001009	208.44	214.49	0.7010
75	0.001023	312.64	318.78	1.0119
100	0.001040	417.29	423.53	1.3024
125	0.001062	522.68	529.05	1.5761
150	0.001087	629.16	635.68	1.8358
175	0.001117	737.16	743.86	2.0842
200	0.001152	847.30	854.22	2.3238
225	0.001195	960.50	967.67	2.5574
250	0.001248	1078.2	1085.7	2.7885
255	0.001260	1102.4	1110.0	2.8348
260	0.001273	1127.0	1134.6	2.8812
265	0.001287	1151.9	1159.6	2.9278

P 70 bar, Ts 285.83 degree C				
T	v	u	h	s
0.01	0.000997	0.1168	7.0936	0.0004
25	0.001000	104.30	111.30	0.3654
50	0.001009	208.29	215.36	0.7006
75	0.001023	312.43	319.59	1.0112
100	0.001040	417.01	424.29	1.3017
125	0.001061	522.32	529.75	1.5752
150	0.001086	628.70	636.30	1.8347
175	0.001116	736.59	744.40	2.0829
200	0.001151	846.58	854.64	2.3223
225	0.001194	959.57	967.92	2.5555
250	0.001246	1076.9	1085.6	2.7861
260	0.001271	1125.6	1134.5	2.8785
270	0.001299	1175.5	1184.6	2.9717
280	0.001331	1227.0	1236.3	3.0661

P 80 bar, Ts 295.01 degree C				
T	v	u	h	s
0.01	0.000996	0.1314	8.101	0.0004
25	0.000999	104.23	112.22	0.3652
50	0.001009	208.15	216.22	0.7001
75	0.001022	312.22	320.40	1.0106
100	0.001039	416.72	425.04	1.3009
125	0.001061	521.96	530.44	1.5743
150	0.001086	628.24	636.93	1.8337
175	0.001115	736.02	744.94	2.0816
200	0.001150	845.86	855.06	2.3207
225	0.001192	958.64	968.18	2.5537
250	0.001245	1075.7	1085.7	2.7837
260	0.001269	1124.2	1134.3	2.8759
270	0.001297	1173.9	1184.3	2.9687
280	0.001328	1225.2	1235.8	3.0627
290	0.001364	1278.4	1289.3	3.1586

P 100 bar, Ts 311.00 degree C				
T	v	u	h	s
0.01	0.000995	0.1591	10.111	0.0005
25	0.000999	104.07	114.06	0.3646
50	0.001008	207.86	217.93	0.6992
75	0.001021	311.80	322.01	1.0094
100	0.001038	416.16	426.55	1.2994
125	0.001059	521.24	531.83	1.5724
150	0.001084	627.34	638.18	1.8315
175	0.001113	734.88	746.02	2.0791
200	0.001148	844.44	855.92	2.3177
225	0.001190	956.82	968.72	2.5500
250	0.001241	1073.3	1085.7	2.7791
275	0.001307	1196.0	1209.1	3.0094
280	0.001323	1221.6	1234.8	3.0561
290	0.001357	1274.2	1287.7	3.1510
300	0.001398	1329.1	1343.1	3.2484

	P	125	bar,	Ts	327.82	degree C
T	v	u	h	s
0.01	0.000994	0.1908	12.62	0.0006
25	0.000997	103.89	116.35	0.3640
50	0.001007	207.50	220.08	0.6980
75	0.001020	311.27	324.03	1.0079
100	0.001037	415.47	428.43	1.2975
125	0.001058	520.35	533.57	1.5702
150	0.001083	626.23	639.76	1.8288
175	0.001112	733.49	747.39	2.0759
200	0.001146	842.69	857.01	2.3139
225	0.001187	954.59	969.42	2.5454
250	0.001237	1070.4	1085.8	2.7734
275	0.001301	1192.0	1208.3	3.0019
300	0.001388	1323.1	1340.4	3.2376
325	0.001524	1473.0	1492.1	3.4964

	P	150	bar,	Ts	342.16	degree C
T	v	u	h	s
0.01	0.000993	0.2196	15.11	0.0006
25	0.000996	103.70	118.64	0.3633
50	0.001006	207.14	222.23	0.6969
75	0.001019	310.76	326.04	1.0063
100	0.001036	414.78	430.32	1.2956
125	0.001057	519.47	535.32	1.5679
150	0.001081	625.12	641.34	1.8262
175	0.001110	732.12	748.76	2.0728
200	0.001143	840.96	858.12	2.3102
225	0.001184	952.40	970.16	2.5410
250	0.001233	1067.5	1086.0	2.7679
275	0.001295	1188.1	1207.6	2.9947
300	0.001378	1317.4	1338.1	3.2275
330	0.001539	1495.5	1518.6	3.5343

	P	200	bar,	Ts	365.75	degree C
T	v	u	h	s
0.01	0.000990	0.2680	20.075	0.0006
25	0.000994	103.33	123.21	0.3619
50	0.001003	206.44	226.51	0.6946
75	0.001017	309.74	330.07	1.0033
100	0.001034	413.43	434.10	1.2918
125	0.001054	517.74	538.82	1.5635
150	0.001078	622.96	644.52	1.8209
175	0.001106	729.43	751.55	2.0667
200	0.001139	837.61	860.39	2.3030
225	0.001178	948.17	971.73	2.5323
250	0.001225	1062.1	1086.6	2.7572
275	0.001284	1180.8	1206.5	2.9810
300	0.001361	1306.9	1334.1	3.2087
325	0.001471	1445.6	1475.0	3.4492
350	0.001665	1612.7	1646.0	3.7288

	P	210	bar,	Ts	369.83	degree C
T	v	u	h	s
0.01	0.000990	0.2763	21.064	0.0006
25	0.000994	103.25	124.12	0.3616
50	0.001003	206.30	227.36	0.6941
75	0.001016	309.54	330.88	1.0027
100	0.001033	413.16	434.86	1.2911
125	0.001053	517.40	539.53	1.5626
150	0.001077	622.54	645.16	1.8199
175	0.001105	728.90	752.11	2.0654
200	0.001138	836.95	860.85	2.3015
225	0.001177	947.34	972.06	2.5306
250	0.001224	1061.0	1086.7	2.7551
275	0.001282	1179.4	1206.3	2.9784
300	0.001358	1304.9	1333.5	3.2051
325	0.001465	1442.5	1473.3	3.4438
350	0.001649	1606.0	1640.7	3.7177

Property Diagrams

The phases of a substance and the relationships between thermodynamic properties are commonly shown on property *diagrams*. Diagrams can be prepared using the property data in these tables. The diagrams are not good for calculations, since one cannot read accurately from a chart, but they are great for a conceptual understanding of various thermodynamic processes. There are five properties of a substance that are usually shown on a property diagrams. These are - pressure P, temperature T, specific volume, v, specific enthalpy, h, and specific entropy, s. When a mixture of two phases, such as water and steam is involved, a sixth property, quality or dryness fraction x, is also used. There are six different types of commonly encountered property diagrams used to depict the properties of saturated steam. These are: Pressure- Temperature (P-T) diagrams, Pressure-Specific Volume (P-v) diagrams, Pressure-Enthalpy (P-h) diagrams, Enthalpy-Temperature (h-T) diagrams, Temperature-entropy (T-s) diagrams, and Enthalpy-Entropy (h-s) or Mollier diagrams. Properties of both saturated liquid and saturated vapor are shown on the same diagram. Therefore, regions demarcating liquid, vapor or a mixture can be identified on such diagrams.

The plotting order of the axis may also be reversed. Do become familiar with the general shape of these diagrams and learn to draw different processes on these diagrams. Different diagrams are useful for different processes. For example the T-s diagram is commonly employed to depict steam cycles such as the Carnot cycle and the Rankine power cycle. The adjoining figures show the T-s and h–s diagram for saturated steam based on the data presented in tables.

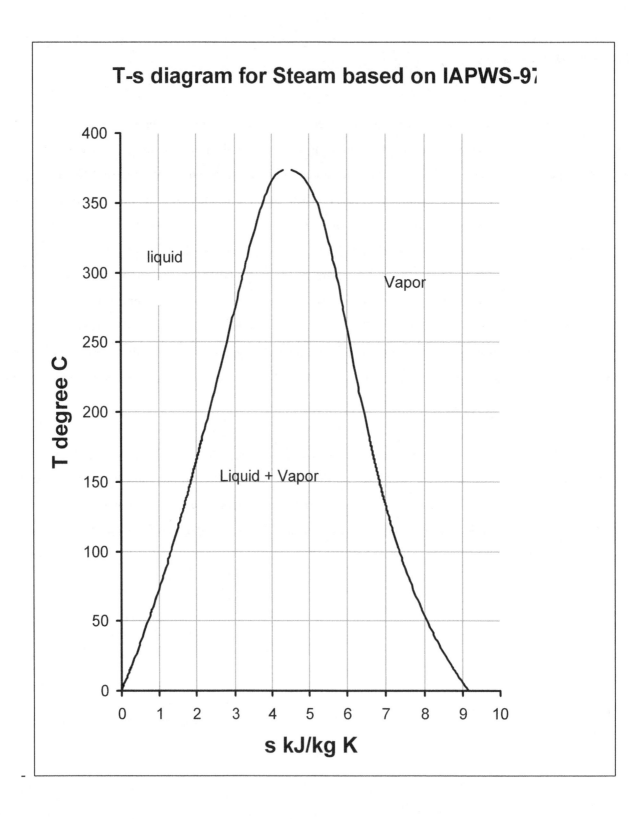

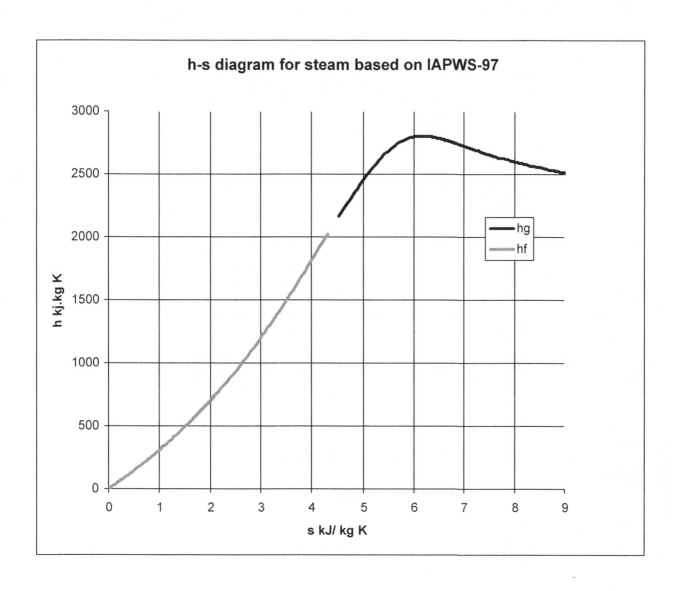

Transport Properties of Water and Steam

The following tables list the most frequently required transport properties of water namely - the density, ρ, specific heat at constant pressure, Cp, Viscosity, μ and the thermal conductivity, k. The Prandtl number, Pr, is a non-dimensional parameter frequently required in heat transfer studies. Therefore, it too has been listed in the tables for easy reference. Another parameter, Turbulent Prandtl number required in computational fluid dynamics (CFD). It can be ascertained from the given values of Prandtl number[4]

Transport Properties of Water (liquid) at 1.01325 Bar (1 atm)

T ^{0} C	ρ kg/ m^3	Cp kJ/kg K	μ Pa s	k W / m K	Pr
0.01	999.84	4.2194	0.001791	0.5611	13.47
1	999.90	4.2160	0.001731	0.5630	12.96
2	999.94	4.2129	0.001673	0.5649	12.48
3	999.97	4.2100	0.001619	0.5668	12.03
4	999.98	4.2074	0.001567	0.5687	11.60
5	999.97	4.2049	0.001518	0.5706	11.19
6	999.94	4.2027	0.001471	0.5725	10.80
7	999.90	4.2007	0.001427	0.5744	10.44
8	999.85	4.1988	0.001385	0.5763	10.09
9	999.78	4.1970	0.001344	0.5782	9.759
10	999.70	4.1954	0.001306	0.5800	9.445
11	999.61	4.1940	0.001269	0.5819	9.147
12	999.50	4.1926	0.001234	0.5838	8.862
13	999.38	4.1913	0.001200	0.5857	8.591
14	999.25	4.1902	0.001168	0.5875	8.332
15	999.10	4.1891	0.001138	0.5894	8.085
16	998.94	4.1881	0.001108	0.5912	7.849
17	998.78	4.1872	0.001080	0.5930	7.624

[4] *Turbulent Prandtl Number in Circular Pipes* by A. Malhotra and S.S. Kang. Int. J. Heat and Mass Transfer Vol. 27, p2158, 1984

T ^{0} C	ρ kg/ m^3	Cp kJ/kg K	μ Pa s	k W / m K	Pr
18	998.60	4.1863	0.001053	0.5949	7.408
19	998.41	4.1855	0.001027	0.5967	7.202
20	998.21	4.1848	0.001002	0.5985	7.004
21	997.99	4.1841	0.0009776	0.6002	6.814
22	997.77	4.1835	0.0009544	0.6020	6.633
23	997.54	4.1829	0.0009322	0.6037	6.458
24	997.30	4.1824	0.0009107	0.6055	6.291
25	997.05	4.1819	0.0008901	0.6072	6.130
26	996.79	4.1814	0.0008702	0.6089	5.976
27	996.52	4.1810	0.0008510	0.6106	5.827
28	996.24	4.1807	0.0008325	0.6122	5.685
29	995.95	4.1803	0.0008146	0.6139	5.547
30	995.65	4.1800	0.0007973	0.6155	5.415
31	995.35	4.1797	0.0007807	0.6171	5.288
32	995.03	4.1795	0.0007646	0.6187	5.165
33	994.71	4.1793	0.0007490	0.6203	5.047
34	994.38	4.1791	0.0007339	0.6218	4.932
35	994.04	4.1789	0.0007193	0.6233	4.822
36	993.69	4.1788	0.0007052	0.6248	4.716
37	993.34	4.1787	0.0006915	0.6263	4.614
38	992.97	4.1786	0.0006783	0.6278	4.515
39	992.60	4.1786	0.0006654	0.6292	4.419
40	992.22	4.1786	0.0006530	0.6306	4.327
41	991.84	4.1785	0.0006409	0.6320	4.237
42	991.45	4.1786	0.0006292	0.6334	4.151
43	991.05	4.1786	0.0006178	0.6348	4.067
44	990.64	4.1787	0.0006068	0.6361	3.986
45	990.22	4.1788	0.0005961	0.6374	3.908
47	989.37	4.1790	0.0005755	0.6399	3.758
48	988.94	4.1792	0.0005657	0.6412	3.687

T ^{0} C	ρ kg/ m^3	Cp kJ/kg K	μ Pa s	k W / m K	Pr
49	988.50	4.1794	0.0005561	0.6424	3.618
50	988.05	4.1796	0.0005469	0.6436	3.551
51	987.59	4.1798	0.0005378	0.6448	3.486
52	987.13	4.1800	0.0005290	0.6459	3.423
53	986.66	4.1803	0.0005204	0.6471	3.362
54	986.19	4.1806	0.0005121	0.6482	3.303
55	985.71	4.1809	0.0005040	0.6493	3.245
56	985.22	4.1812	0.0004961	0.6503	3.189
57	984.73	4.1816	0.0004884	0.6514	3.135
58	984.23	4.1820	0.0004809	0.6524	3.082
59	983.72	4.1823	0.0004735	0.6534	3.031
60	983.21	4.1828	0.0004664	0.6544	2.981
61	982.69	4.1832	0.0004594	0.6554	2.933
62	982.17	4.1837	0.0004527	0.6563	2.885
63	981.64	4.1841	0.0004460	0.6572	2.840
64	981.11	4.1846	0.0004396	0.6581	2.795
65	980.57	4.1852	0.0004333	0.6590	2.752
66	980.02	4.1857	0.0004271	0.6599	2.709
67	979.47	4.1863	0.0004211	0.6607	2.668
68	978.91	4.1869	0.0004152	0.6616	2.628
69	978.35	4.1875	0.0004095	0.6624	2.589
70	977.78	4.1881	0.0004039	0.6631	2.551
71	977.21	4.1887	0.0003984	0.6639	2.514
72	976.63	4.1894	0.0003931	0.6647	2.478
73	976.04	4.1901	0.0003879	0.6654	2.442
74	975.45	4.1908	0.0003827	0.6661	2.408
75	974.86	4.1915	0.0003777	0.6668	2.375
76	974.26	4.1923	0.0003729	0.6675	2.342
77	973.65	4.1931	0.0003681	0.6681	2.310
79	972.42	4.1947	0.0003588	0.6694	2.248

T ^{0} C	ρ kg/ m^3	Cp kJ/kg K	μ Pa s	k W / m K	Pr
80	971.80	4.1955	0.0003544	0.6700	2.219
81	971.18	4.1964	0.0003500	0.6706	2.190
82	970.55	4.1972	0.0003457	0.6712	2.162
83	969.91	4.1981	0.0003415	0.6718	2.134
84	969.27	4.1991	0.0003374	0.6723	2.107
85	968.62	4.2000	0.0003333	0.6728	2.081
86	967.97	4.2010	0.0003294	0.6734	2.055
87	967.32	4.2019	0.0003255	0.6739	2.030
88	966.65	4.2030	0.0003218	0.6743	2.005
89	965.99	4.2040	0.0003180	0.6748	1.981
90	965.32	4.2050	0.0003144	0.6753	1.958
91	964.64	4.2061	0.0003109	0.6757	1.935
92	963.96	4.2072	0.0003074	0.6762	1.912
93	963.28	4.2083	0.0003039	0.6766	1.890
94	962.59	4.2094	0.0003006	0.6770	1.869
95	961.90	4.2106	0.0002973	0.6774	1.848
96	961.20	4.2117	0.0002941	0.6777	1.827
97	960.49	4.2129	0.0002909	0.6781	1.807
98	959.78	4.2141	0.0002878	0.6784	1.788
99	959.07	4.2154	0.0002847	0.6788	1.768
99.5	958.71	4.2160	0.0002832	0.6789	1.759

Transport Properties of Water (vapor) at 1.01325 Bar (1 atm)

T ^{0} C	ρ kg/ m^3	Cp kJ/kg K	μ Pa s	k W / m K	Pr
100	0.5976	2.0773	0.00001227	0.02510	1.0156
125	0.5576	2.0123	0.00001321	0.02688	0.9892
150	0.5232	1.9867	0.00001418	0.02886	0.9761
175	0.4931	1.9769	0.00001517	0.03101	0.9672
200	0.4664	1.9762	0.00001618	0.03329	0.9603
225	0.4426	1.9810	0.00001719	0.03568	0.9545
250	0.4211	1.9894	0.00001822	0.03818	0.9494
275	0.4017	2.0000	0.00001925	0.04076	0.9447
300	0.3840	2.0122	0.00002029	0.04343	0.9402
325	0.3678	2.0255	0.00002133	0.04616	0.9359
350	0.3529	2.0397	0.00002237	0.04897	0.9318
375	0.3392	2.0545	0.00002341	0.05184	0.9279
400	0.3266	2.0697	0.00002445	0.05476	0.9241
425	0.3148	2.0855	0.00002549	0.05774	0.9205
450	0.3039	2.1015	0.00002652	0.06077	0.9171
475	0.2937	2.1179	0.00002755	0.06385	0.9138
500	0.2842	2.1346	0.00002857	0.06698	0.9107
525	0.2753	2.1515	0.00002959	0.07015	0.9077
550	0.2669	2.1686	0.00003061	0.07336	0.9048
575	0.2590	2.1858	0.00003162	0.07661	0.9020
600	0.2516	2.2032	0.00003262	0.07990	0.8994
625	0.2445	2.2206	0.00003361	0.08322	0.8969
650	0.2379	2.2381	0.00003460	0.08658	0.8945
675	0.2316	2.2557	0.00003558	0.08996	0.8921
700	0.2257	2.2732	0.00003655	0.09338	0.8899
725	0.2200	2.2908	0.00003752	0.09683	0.8877
750	0.2146	2.3083	0.00003848	0.10031	0.8855
775	0.2095	2.3258	0.00003943	0.10381	0.8835
800	0.2046	2.3434	0.00004038	0.10733	0.8815

Density Inversion of Water

The preceding data table shows the density inversion that water undergoes at 4 ^{0}C. This density inversion is also shown in the following chart

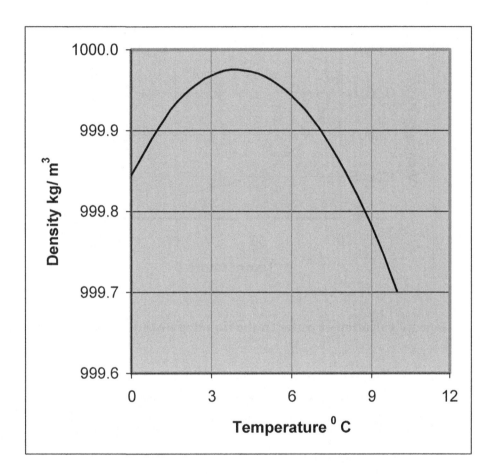

Density of water in the vicinity of 4 ^{0}C at 1 Bar Pressure

Viscosity Variations with Temperature

The preceding tables for Transport Properties of Water at 1.01325 Bar (1atm) illustrate that the dynamic viscosity of liquid water decreases with an increase of temperature whereas in the vapor state the dynamic viscosity increases with an increase of temperature. The variations are illustrated in figure 2 and figure 3.

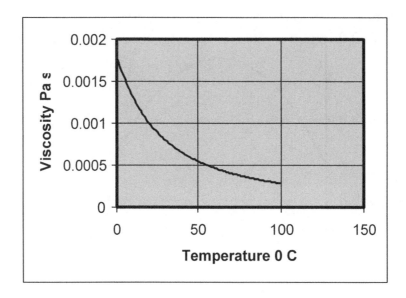

Viscosity variation of water in the liquid state at P = 1.01325 Bar

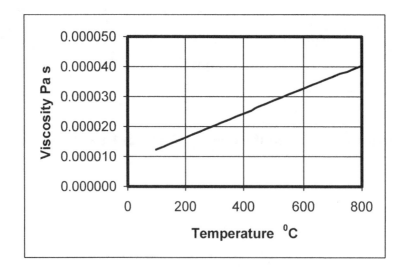

Viscosity variation of water in the vapor state at P = 1.01325 Bar

Transport Properties of Liquid Water at different pressures

P Bar	T ^{0}C	ρ kg/ m^3	Cp kJ/kg K	μ Pa s	k W / m K	Pr
0.04	25	997.00	4.1822	0.0008901	0.6072	6.13
1	25	997.05	4.1819	0.0008901	0.6072	6.13
1.1	25	997.05	4.1819	0.0008901	0.6072	6.13
1.2	25	997.06	4.1818	0.0008901	0.6072	6.13
1.3	25	997.06	4.1818	0.0008901	0.6072	6.13
1.4	25	997.07	4.1818	0.0008901	0.6072	6.13
1.5	25	997.07	4.1818	0.0008901	0.6072	6.13
1.6	25	997.07	4.1817	0.0008901	0.6072	6.13
1.7	25	997.08	4.1817	0.0008901	0.6072	6.13
1.8	25	997.08	4.1817	0.0008901	0.6072	6.13
1.9	25	997.09	4.1816	0.0008901	0.6072	6.13
2	25	997.09	4.1816	0.0008901	0.6072	6.13
3	25	997.14	4.1813	0.00089	0.6073	6.13
4	25	997.18	4.1810	0.00089	0.6073	6.13
5	25	997.23	4.1807	0.00089	0.6074	6.13
6	25	997.27	4.1805	0.00089	0.6074	6.13
7	25	997.32	4.1802	0.0008899	0.6075	6.12
8	25	997.36	4.1799	0.0008899	0.6075	6.12
9	25	997.41	4.1796	0.0008899	0.6076	6.12
10	25	997.45	4.1793	0.0008899	0.6076	6.12
15	25	997.68	4.1779	0.0008898	0.6078	6.12
20	25	997.90	4.1764	0.0008896	0.6081	6.11
25	25	998.13	4.1750	0.0008895	0.6083	6.11
30	25	998.35	4.1736	0.0008894	0.6085	6.10
35	25	998.57	4.1722	0.0008893	0.6087	6.10
40	25	998.80	4.1708	0.0008892	0.6090	6.09
50	25	999.24	4.1680	0.000889	0.6094	6.08
55	25	999.47	4.1666	0.0008889	0.6096	6.08

40

--- Steam Property Tables ---

P Bar	T ^{0} C	ρ kg/m^3	Cp kJ/kg K	μ Pa s	k W/m K	Pr
60	25	999.69	4.1652	0.0008888	0.6099	6.07
65	25	999.91	4.1638	0.0008887	0.6101	6.07
70	25	1000.1	4.1625	0.0008886	0.6103	6.06
75	25	1000.4	4.1611	0.0008885	0.6105	6.06
80	25	1000.6	4.1597	0.0008884	0.6108	6.05
85	25	1000.8	4.1584	0.0008883	0.6110	6.05
90	25	1001.0	4.1570	0.0008882	0.6112	6.04
95	25	1001.2	4.1557	0.0008881	0.6115	6.04
100	25	1001.5	4.1543	0.0008880	0.6117	6.03
105	25	1001.7	4.1530	0.0008879	0.6119	6.03
110	25	1001.9	4.1517	0.0008878	0.6121	6.02
115	25	1002.1	4.1503	0.0008877	0.6124	6.02
120	25	1002.3	4.1490	0.0008876	0.6126	6.01
125	25	1002.6	4.1477	0.0008876	0.6128	6.01
130	25	1002.8	4.1464	0.0008875	0.6130	6.00
140	25	1003.2	4.1438	0.0008873	0.6135	5.99
150	25	1003.7	4.1412	0.0008871	0.6140	5.98
160	25	1004.1	4.1386	0.0008870	0.6144	5.97
170	25	1004.5	4.1360	0.0008868	0.6149	5.97
180	25	1005.0	4.1335	0.0008867	0.6153	5.96
190	25	1005.4	4.1310	0.0008865	0.6158	5.95
200	25	1005.8	4.1285	0.0008864	0.6162	5.94
210	25	1006.3	4.1260	0.0008863	0.6167	5.93
220	25	1006.7	4.1235	0.0008862	0.6172	5.92
230	25	1007.1	4.1211	0.000886	0.6176	5.91
240	25	1007.6	4.1186	0.0008859	0.6181	5.90
250	25	1008.0	4.1162	0.0008858	0.6185	5.89
260	25	1008.4	4.1138	0.0008857	0.6190	5.89
270	25	1008.8	4.1115	0.0008856	0.6194	5.88
280	25	1009.3	4.1091	0.0008855	0.6199	5.87
300	25	1010.1	4.1044	0.0008853	0.6208	5.85

P Bar	T ^{0} C	ρ kg/ m^3	Cp kJ/kg K	μ Pa s	k W / m K	Pr
310	25	1010.5	4.1021	0.0008852	0.6213	5.85
320	25	1011.0	4.0999	0.0008851	0.6217	5.84
330	25	1011.4	4.0976	0.0008851	0.6222	5.83
340	25	1011.8	4.0953	0.0008850	0.6226	5.82
350	25	1012.2	4.0931	0.0008849	0.6231	5.81
360	25	1012.6	4.0909	0.0008849	0.6235	5.81
370	25	1013.1	4.0887	0.0008848	0.6240	5.80
380	25	1013.5	4.0865	0.0008848	0.6244	5.79
390	25	1013.9	4.0843	0.0008847	0.6249	5.78
400	25	1014.3	4.0821	0.0008847	0.6253	5.78
410	25	1014.7	4.0800	0.0008846	0.6258	5.77
420	25	1015.1	4.0779	0.0008846	0.6262	5.76
430	25	1015.6	4.0758	0.0008846	0.6267	5.75
440	25	1016.0	4.0737	0.0008845	0.6271	5.75
450	25	1016.4	4.0716	0.0008845	0.6276	5.74
460	25	1016.8	4.0695	0.0008845	0.6280	5.73
470	25	1017.2	4.0675	0.0008845	0.6285	5.72
480	25	1017.6	4.0654	0.0008845	0.6289	5.72
490	25	1018.0	4.0634	0.0008845	0.6294	5.71
500	25	1018.4	4.0614	0.0008845	0.6298	5.70
510	25	1018.8	4.0594	0.0008845	0.6303	5.70
520	25	1019.3	4.0574	0.0008845	0.6307	5.69
530	25	1019.7	4.0555	0.0008845	0.6312	5.68
540	25	1020.1	4.0535	0.0008845	0.6316	5.68
550	25	1020.5	4.0516	0.0008846	0.6321	5.67
560	25	1020.9	4.0497	0.0008846	0.6325	5.66
570	25	1021.3	4.0478	0.0008846	0.6330	5.66
580	25	1021.7	4.0459	0.0008847	0.6334	5.65
590	25	1022.1	4.0440	0.0008847	0.6339	5.64
600	25	1022.5	4.0422	0.0008847	0.6343	5.64
620	25	1023.3	4.0385	0.0008848	0.6352	5.63

P Bar	T ^{0}C	ρ kg/ m^3	Cp kJ/kg K	μ Pa s	k W / m K	Pr
630	25	1023.7	4.0366	0.0008849	0.6356	5.62
640	25	1024.1	4.0348	0.0008849	0.6361	5.61
650	25	1024.5	4.0330	0.000885	0.6365	5.61
660	25	1024.9	4.0312	0.0008851	0.6369	5.60
670	25	1025.3	4.0295	0.0008852	0.6374	5.60
680	25	1025.6	4.0277	0.0008852	0.6378	5.59
690	25	1026.0	4.0260	0.0008853	0.6383	5.58
700	25	1026.4	4.0242	0.0008854	0.6387	5.58
710	25	1026.8	4.0225	0.0008855	0.6391	5.57
720	25	1027.2	4.0208	0.0008856	0.6396	5.57
730	25	1027.6	4.0191	0.0008857	0.6400	5.56
740	25	1028.0	4.0174	0.0008858	0.6404	5.56
750	25	1028.4	4.0158	0.0008859	0.6409	5.55
760	25	1028.8	4.0141	0.000886	0.6413	5.55
770	25	1029.2	4.0124	0.0008861	0.6417	5.54
780	25	1029.5	4.0108	0.0008862	0.6422	5.54
790	25	1029.9	4.0092	0.0008863	0.6426	5.53
800	25	1030.3	4.0076	0.0008864	0.6430	5.52
810	25	1030.7	4.0060	0.0008866	0.6434	5.52
820	25	1031.1	4.0044	0.0008867	0.6439	5.51
830	25	1031.5	4.0028	0.0008868	0.6443	5.51
840	25	1031.9	4.0012	0.000887	0.6447	5.50
850	25	1032.2	3.9997	0.0008871	0.6451	5.50

Transport properties of water Vapor at 400 degree C and different pressures

P Bar	T ^{0} C	ρ kg/ m^3	Cp kJ/kg K	μ Pa s	k W / m K	Pr
0.04	400	0.013	2.0637	0.00002445	0.0547	0.92
1	400	0.322	2.0697	0.00002445	0.0548	0.92
1.2	400	0.387	2.0709	0.00002445	0.0548	0.92
1.3	400	0.419	2.0715	0.00002445	0.0548	0.92
1.4	400	0.451	2.0722	0.00002445	0.0548	0.92
1.5	400	0.484	2.0728	0.00002445	0.0548	0.92
1.6	400	0.516	2.0734	0.00002445	0.0548	0.92
1.7	400	0.548	2.0741	0.00002445	0.0548	0.92
1.8	400	0.581	2.0747	0.00002445	0.0548	0.92
1.9	400	0.613	2.0753	0.00002445	0.0548	0.93
2	400	0.645	2.0760	0.00002445	0.0549	0.93
3	400	0.969	2.0823	0.00002444	0.0549	0.93
4	400	1.294	2.0887	0.00002444	0.0550	0.93
5	400	1.620	2.0952	0.00002444	0.0551	0.93
6	400	1.947	2.1017	0.00002444	0.0552	0.93
7	400	2.274	2.1083	0.00002443	0.0553	0.93
8	400	2.602	2.1149	0.00002443	0.0554	0.93
9	400	2.932	2.1216	0.00002443	0.0555	0.93
10	400	3.262	2.1284	0.00002442	0.0556	0.93
15	400	4.926	2.1632	0.00002441	0.0561	0.94
20	400	6.613	2.1997	0.00002440	0.0566	0.95
25	400	8.325	2.2379	0.00002439	0.0572	0.96
30	400	10.06	2.2780	0.00002438	0.0577	0.96
35	400	11.84	2.3201	0.00002438	0.0583	0.97
40	400	13.62	2.3642	0.00002437	0.0588	0.98
45	400	15.44	2.4105	0.00002437	0.0594	0.99
50	400	17.29	2.4590	0.00002437	0.0601	**1.00**

44

P Bar	T ^{0} C	ρ kg/ m^3	Cp kJ/kg K	μ Pa s	k W / m K	Pr
60	400	21.09	2.5632	0.00002437	0.0614	1.02
65	400	23.04	2.6191	0.00002438	0.0621	1.03
70	400	25.02	2.6778	0.00002439	0.0628	1.04
75	400	27.05	2.7392	0.00002440	0.0636	1.05
80	400	29.11	2.8037	0.00002441	0.0643	1.06
85	400	31.22	2.8714	0.00002442	0.0652	1.08
90	400	33.37	2.9425	0.00002444	0.0660	1.09
95	400	35.57	3.0172	0.00002446	0.0669	1.10
100	400	37.82	3.0958	0.00002449	0.0679	1.12
105	400	40.12	3.1785	0.00002451	0.0689	1.13
110	400	42.48	3.2658	0.00002454	0.0699	1.15
115	400	44.90	3.3578	0.00002458	0.0710	1.16
120	400	47.38	3.4551	0.00002461	0.0722	1.18
125	400	49.92	3.5581	0.00002465	0.0734	1.20
130	400	52.54	3.6672	0.00002470	0.0747	1.21
140	400	58.00	3.9064	0.00002480	0.0775	1.25
150	400	63.81	4.1778	0.00002493	0.0807	1.29
160	400	70.02	4.4882	0.00002508	0.0843	1.34
170	400	76.70	4.8463	0.00002526	0.0885	1.38
180	400	83.93	5.2645	0.00002547	0.0932	1.44
190	400	91.82	5.7605	0.00002573	0.0988	1.50
200	400	100.5	6.3601	0.00002603	0.1055	1.57
210	400	110.2	7.1006	0.00002641	0.1134	1.65
220	400	121.1	8.0332	0.00002687	0.1230	1.76
230	400	133.7	9.2285	0.00002745	0.1347	1.88
240	400	148.6	10.803	0.00002819	0.1494	2.04
250	400	166.5	13.002	0.00002917	0.1681	2.26
260	400	189.2	16.237	0.00003053	0.1928	2.57
270	400	219.0	21.142	0.00003252	0.2259	3.04
280	400	259.4	27.310	0.00003551	0.2674	3.63
290	400	309.5	29.847	0.00003964	0.3064	3.86

P Bar	T ^{0} C	ρ kg/ m^3	Cp kJ/kg K	μ Pa s	k W / m K	Pr
310	400	394.9	20.624	0.00004751	0.3472	2.82
320	400	422.5	16.932	0.00005027	0.3589	2.37
330	400	443.7	14.496	0.00005246	0.3688	2.06
340	400	460.7	12.837	0.00005426	0.3773	1.85
350	400	474.9	11.650	0.00005579	0.3849	1.69
360	400	487.1	10.763	0.00005712	0.3918	1.57
370	400	497.7	10.075	0.00005831	0.3980	1.48
380	400	507.1	9.5262	0.00005938	0.4038	1.40
390	400	515.6	9.0766	0.00006037	0.4091	1.34
400	400	523.4	8.7012	0.00006127	0.4140	1.29
410	400	530.5	8.3826	0.00006212	0.4186	1.24
420	400	537.1	8.1082	0.00006291	0.4230	1.21
430	400	543.3	7.8692	0.00006365	0.4271	1.17
440	400	549.0	7.6588	0.00006436	0.4310	1.14
450	400	554.5	7.4719	0.00006503	0.4347	1.12
460	400	559.6	7.3046	0.00006567	0.4383	1.09
470	400	564.5	7.1538	0.00006628	0.4417	1.07
480	400	569.1	7.0171	0.00006687	0.4450	1.05
490	400	573.5	6.8924	0.00006744	0.4481	1.04
500	400	577.7	6.7781	0.00006798	0.4512	1.02
510	400	581.8	6.6729	0.00006851	0.4541	1.01
520	400	585.7	6.5757	0.00006902	0.4570	0.99
530	400	589.4	6.4855	0.00006952	0.4598	0.98
540	400	593.0	6.4016	0.00007000	0.4624	0.97
550	400	596.5	6.3232	0.00007046	0.4651	0.96
560	400	599.9	6.2499	0.00007092	0.4676	0.95
570	400	603.2	6.1810	0.00007136	0.4701	0.94
580	400	606.3	6.1163	0.00007180	0.4726	0.93
590	400	609.4	6.0552	0.00007222	0.4750	0.92
600	400	612.4	5.9975	0.00007263	0.4773	0.91
610	400	615.3	5.9428	0.00007304	0.4796	0.91

P Bar	T ^{0} C	ρ kg/ m^3	Cp kJ/kg K	μ Pa s	k W / m K	Pr
630	400	620.9	5.8418	0.00007382	0.4840	0.89
640	400	623.6	5.7949	0.00007420	0.4862	0.88
650	400	626.2	5.7502	0.00007457	0.4883	0.88
660	400	628.7	5.7076	0.00007494	0.4904	0.87
670	400	631.2	5.6668	0.00007530	0.4925	0.87
680	400	633.7	5.6278	0.00007565	0.4945	0.86
690	400	636.1	5.5905	0.00007600	0.4965	0.86
700	400	638.4	5.5547	0.00007634	0.4985	0.85
710	400	640.7	5.5203	0.00007668	0.5004	0.85
720	400	643.0	5.4872	0.00007701	0.5023	0.84
730	400	645.2	5.4554	0.00007733	0.5042	0.84
740	400	647.3	5.4248	0.00007765	0.5061	0.83
750	400	649.4	5.3953	0.00007797	0.5079	0.83
760	400	651.5	5.3668	0.00007828	0.5097	0.82
770	400	653.6	5.3393	0.00007859	0.5115	0.82
780	400	655.6	5.3128	0.00007889	0.5133	0.82
790	400	657.5	5.2871	0.00007919	0.5150	0.81
800	400	659.5	5.2623	0.00007949	0.5168	0.81
810	400	661.4	5.2382	0.00007978	0.5185	0.81
820	400	663.3	5.2149	0.00008007	0.5202	0.80
830	400	665.1	5.1924	0.00008036	0.5219	0.80
840	400	667.0	5.1705	0.00008064	0.5235	0.80
850	400	668.8	5.1492	0.00008092	0.5252	0.79

Made in United States
Orlando, FL
24 October 2024